Laboratory Fume Hoods

Laboratory Fume Hoods

A User's Manual

G. Thomas Saunders

A WILEY-INTERSCIENCE PUBLICATION

John Wiley & Sons, Inc.

NEW YORK / CHICHESTER / BRISBANE / TORONTO / SINGAPORE

Library of Congress Cataloging in Publication Data:

Saunders, G. Thomas, 1924-
    Laboratory fume hoods : a users manual / G. Thomas Saunders.
        p. cm.
    Includes bibliographical references and index.
    ISBN 0-471-56935-6 (alk. paper)
    1. Fume hoods. I. Title.
QD54.F85S38 1993
542'.1--dc20   92-39024
                                              CIP

Printed in the United States of America

10 9 8 7 6 5 4 3 2 1

Preface

American consumers have a wide variety of published product evaluations available to them prior to making any type of purchase. You can find "good–better–best" comments covering most items from automobiles to zucchini. Not all buyers follow these guides but the data are there and are usually presented with reasonable and viable details. Manufacturers obviously read the critiques; if a product exhibits a flaw in a January report it is usually corrected by June.

In the world of laboratory fume hoods, this is not the case. The potential buyer of hoods, be it an architect, engineer, researcher, or purchasing manager, is left to digest catalogs and then be subjected to lengthy sales presentations. Since a manufacturer never says a bad thing about his own product and most salespeople know little or nothing about fume hoods, the buyer must proceed with the old "byguess & bygosh" method.

Hopefully, it will be possible to start a reversal of this protocol. We are going to dissect the fume hood down to its very bones and give the reasons as to how and why a fume hood works. We will test, locate, ventilate, and maintain hoods, both old and new.

Knowing our goal I do hope that you as the designer, user, or buyer will install safe fume hoods, or upgrade existing hoods, in your facility. With luck the manufacturers, planners, and engineers will also read and heed what is said and produce a safer fume hood and a more compatible fume hood system. If all of this happens, this book will be a huge success.

G. THOMAS SAUNDERS

Durham, North Carolina

Acknowledgments

From Dr. Jay Young, consultant from Silver Spring, Maryland, I received the intitial push to even consider writing this book. Jay was a constant "peer reader" and very welcome advisor during this year-long process. He was my "executive conscience." Dr. Malcolm Renfrew, retired Chairman of the Department of Chemistry and Professor Emeritus at the University of Idaho, was always available with advice and he was my "technical conscience." Mr. Thomas Smith, MSEE of Contamination Control Technologies, Raleigh, North Carolina, assisted me in the area of hood field testing and in helping with the illustrations; he was my "test conscience." My good wife, Barbara, and youngest son, Stephen, were my "time consciences" demanding that I share my ham radio time with writing the book.

To all these wonderful people I owe a deep sense of gratitude especially now that this project has been completed.

G. T. S.

Contents

Acronyms

ACGIH	American Conference of Governmental Industrial Hygienists
AEC	Atomic Energy Commission
ASHRAE	American Society of Heating, Refrigerating and Air-Conditioning Engineers
CCC	Concentrate, Confine, and Control
EPA	Environmental Protection Agency
HEPA	High-Efficiency Particulate Arrestor
HVAC	Heating, Ventilating and Air Conditioning
OSHA	Occupational Health and Safety Administration
SAMA	Scientific Apparatus Makers Association
VAV	Variable Air Volume

Introduction

Fume hoods are not complicated; they operate on some basic parameters. Air is lazy: it goes where directed and rarely thinks for itself. A fume hood, per se, is only one part of a very demanding system. Keeping this in mind, let us see how the fume hood developed and how far it has come through these many years.

The first fume hood was the fireplace used by an alchemist. If you compare the profile of a fireplace to that of a fume hood, you see a great deal of similarity. The fireplace had a fairly tall chimney. The stack height, the thermal gradations caused by the fire, and the aspirating effect of the outside wind conditions could create a respectable draft. To increase the draft, early day ventilation engineers (mid-1800s) added gas-burning rings in the stack to achieve greater thermal lift. During the Industrial Revolution the gas ring gave way to a mechanical fan. At about this point we started to see defined laboratories with honest-to-gosh fume hoods in lieu of the old fireplace.

Little by little other niceties were added. The front sash was originally a hinged door and most likely added by the English since even today a British scientist refers to the fume hood as a fume cupboard. Various other services were added as they became available and equipment requirements demanded them (e.g., air, gas, electrical outlets, water, etc).

The first major improvement in the development of the fume hood proper was the addition of the back baffle system. This slotted panel forced the air to be taken from the area of the hood working surface as well as from the top canopy area. With this addition fume hoods really started to work as a safety device—a device that protected the user by at least qualitatively keeping fumes from his or her breathing zone. In the 1940s the Harvard School of Public Health was given a contract by the (then) Atomic Energy Commission to develop equipment items concerned with fume hood operation and safety. This task force was headed by Dr. Leslie Silverman, a truly wonderful man and a most imposing scientist. His team fostered, among other things, the development of the HEPA (high-efficiency particulate arrestor) filter and the streamline shaped entrance for the fume hood. By 1950 we had a great particulate filter and a fume hood that was head and shoulders above anything that had been produced up to that time.

Despite the constant claims of the various hood manufacturers little has been added to basic hood design since then. In the 1960s, 1970s and 1980s a lot of attention was given to auxiliary air fume hoods and some of these devices worked quite well. As a price for their excellence, they introduced a wide spectrum of balance and maintenance problems, and with the introduction of the variable air volume (VAV) systems in the last few years the auxiliary air hood has lost much of its glitter.

When fume hoods became a larger and more significant part of every laboratory, there arose a demand by the fume hood user to know if the hood was indeed working as a safety device. The easiest parameter to explore and to measure was the velocity of the air entering the front of the fume hood, the face velocity.

In the 1940s and 1950s a face velocity of 50 feet per minute (fpm) was considered adequate. As the fume hood population doubled and doubled and doubled again, there was a group of "experts" that seemed preordained to improve hood safety by increasing the face velocity. The value went from 50 to 150 fpm and would still be going up and up and up if the American Society of Heating, Refrigeration and Air Conditioning Engineers (ASHRAE) had not sponsored a most significant research project in the mid-1970s. K. J. Caplan and G. W. Knutson, then at the University of Minnesota, undertook a project to find a quantitative method to test fume hoods. They steered clear of the smoke and mirrors that had been used up to that time. They were most successful in developing a quantitative measure of hood performance.

Their research provided the laboratory community with a procedure for establishing fume hood performance—or lack of same. Their findings were published in 1978 with the release of ASHRAE RP-70 (1) and is the basis for ASHRAE Standard 110-1985 (2). For the first time face velocities were relegated to their proper place in the safety chain, and the safe spread was reduced to a 60- to 100-fpm range. The sensitivity of the testing was increased to 0.01 ppm from the 300-ppm range of earlier smoke tests. Testing became objective, not subjective, and was reproducible by any number of trained people given the proper equipment.

That's where we are today. No big design changes have been introduced since the 1940s, but now we have a quantitative method of evaluation. In this book we go back to the 1940s and show what design can do for us when properly applied, and then we show the "good–better–best" in fume hood performance. The performance evaluations, which are all on-site and field-generated, will give food for thought for the architect, engineer, and user. Since a fume hood is but one part of a complicated system, safe fume hood performance depends on the proper design and discipline of all parties.

It is my hope to make all of you fume hood experts. Many people today profess to be fume hood experts, but it is my honest opinion that there are darned few that should be in this classification. As you read this book and question some of the professed experts you will understand why. If you get the feeling that I am also disenchanted with hood manufacturers, you are correct. A couple of the hood producers have competent designers, the others may not even be marginal. Considering that there are six major and three minor companies in the field, the odds of you getting real professional help from a manufacturer are slim.

REFERENCES

1. K. J. Caplan and G. W. Knutson, "Laboratory Fume Hoods, a Performance Test," RP-70, *ASHRAE Trans.*, 84, (I)(1978).
2. Method of Testing Laboratory Fume Hoods, ASHRAE Standard 110-1985, American Society of Heating, Refrigerating and Air Conditioning Engineers, Atlanta, 1985.

Laboratory Fume Hoods

1

Road Map

Consider being in New York and having to drive to Boston. How do you go? The route that you take depends on you and your mission.

If you are pressed for time you head for the fast-track turnpike systems and make the trip without enjoying the beautiful scenery and events that are available. If you are a lover of the ocean, you drive to the end of Long Island, take a ferry to Connecticut, go east to Rhode Island, enjoy a stay in Newport and then proceed to Boston. For the back road enthusiast, you clear the New York City area, travel diagonally across Connecticut and part of Massachusetts, and enjoy the magnificent countryside.

We are going to take a journey to an effective fume hood system but our routes will be decided by a different set of interests.

What is a fume hood? Why is it needed? How does it work? What makes one hood work better than another? How do I know my hood is working safely?

We will take our tour of this defined countryside, question by question, and when we arrive at our destination we hope to be more at ease with our laboratory environs. Hopefully, we will enjoy a healthier journey.

So that we do not get lost in the process, here is a list of the checkpoints for our trek. They are shown here in their descending order of importance.

1. Room air patterns.
2. Basic Hood Design.
3A. Face velocity.
3B. System design.
4. User discipline.

To better understand our trip, let us put these checkpoints into their proper perspective.

Your hood is not performing as you would like. There are odors coming into the laboratory that could only originate in the hood. Where do we start to investigate the cause?

Start with the simplest and most obvious causes. If your hood is cluttered and you are working at the front edge then your discipline could be the cause (Chapter 7). If you clean up your act and you still have some disturbing odor, then you must go one step more. If opening and closing your lab door requires a little extra effort and the room temperature fluctuates a bit more than you find comfortable, then your system design may be the culprit (Chapter 6). If the hood face velocity is between 60 and 100 fpm then increasing the velocity will do you little if any good (Chapter 5). We check and find that the back baffles are not properly adjusted and that the bottom front air foil has been removed so it could be that the basic hood design is also a problem (Chapter 3). We correct these items but we still have a faint odor that we started out to banish from our laboratory. Now lets do a small amount of qualitative testing (Chapter 8). We take a 30-second smoke bomb and discharge it inside the hood and see a very slight, but not continuous, outflow of smoke into the room. Now we have cornered the primary cause: room air patterns. Somehow we have a cross draft that is pulling the hood atmosphere into the room.

Until you correct the major problem area the other items are not as important. Often poor hood performance is not always due to one cause. It can be a collection of a lot of small things, and you have to correct each one until they are all accounted for. That is why it is important to read and absorb all those chapters that hold the key, or keys, to help you in your trip toward better hood performance.

PERSONALIZED ROAD MAPS

As with the trip to Boston there are different routes for different folks. Let us consider those disciplines that are more administrative than user oriented. How do staff personnel use this book for maximum benefit without too many side trips? Here is a partial list of those functions to be considered:

1. The purchasing of new hoods.
2. The planning of new facilities.
3. Designing hood systems for new buildings.
4. Upgrading existing laboratories.
5. Performing safety evaluations of laboratories.

The buyer of new hoods should start with Hood Design (Chapters 3 and 4), and quantitative Testing (Chapter 9), proceed to Specifications (Chapter 10).

For the committee planning a new facility the place to begin would be room air patterns (Chapter 2), followed by system design (Chapter 6), hood design (Chapters 3 and 4), face velocities (Chapter 5) and then quantitative testing (Chapter 9).

The engineer starting to lay out systems for a new project could be best served by starting with face velocities (Chapter 5) followed by room air patterns (Chapter 2), systems design (Chapter 6), and then quantitative testing (Chapter 9).

For upgrading existing facilities a good starting point would be hood design (Chapters 3 and 4), followed by room air patterns (Chapter 2), face velocities (Chapter 5), system design (Chapter 6), and testing (Chapters 8 and 9).

For the health professional contemplating an evaluation of facilities, I would suggest a route that would start with face velocities (Chapter 5), continue on to discipline (Chapter 7), and testing (Chapters 8 and 9), and then finish up with room air patterns (Chapter 2).

Again I repeat: these are the short routes but after these initial excursions into the book take the time to read and digest it all. It will be enlightening.

2

Room Air Patterns

As one of the first steps in the planning of a new laboratory building have the environmental health staff set the primary fume hood face velocity. With this parameter, and the anticipated heat loading per room, the ventilation engineer can calculate the air volumes required for the facility.

At this early point in the planning process the design group will make a decision that may well make or break the integrity of the fume hood system. A simple question is posed with a not so simple answer. How do we get the air volume into the laboratory space with minimal room air turbulence?

The decision makers should have their copy of *Industrial Ventilation Manual*, 17th edition or later, as published by the American Conference of Governmental Industrial Hygienists (ACGIH)[1], opened to VS-204 and VS-204.1. It is here that the acceptable conditions for laboratory room air entry are outlined. Exit discharge velocity from the room air diffusers should not exceed 60 percent of the face velocities assigned to the fume hoods.

This is one of the most important parameters that must be followed by the heating, ventilating, and air conditioning (HVAC) staff. It rules out the use of high-velocity blade diffusers that are used in practically all of today's buildings. A common blade type diffuser is shown in Figure 2.1.

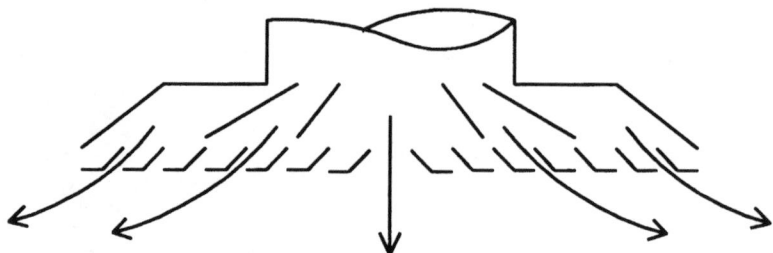

FIGURE 2.1 *Blade type diffuser.*

In this type of air diffuser the average directed velocities vary from 200 to 800 fpm. This is well outside the acceptable range for laboratories. The easiest solution is to not use this type of equipment. There are units that are specifically made to slow down the exit velocity. One such diffuser is shown in Figure 2.2.

These air entry devices are made in both 2-foot by 2-foot and 2-foot by 4-foot modules so that they easily drop into a suspended ceiling grid system. Take the total volume of room exhaust air (cfm) and divide by 60 percent of the selected hood face velocity (fpm). This gives you the square foot of diffuser face required for that particular laboratory. To determine the number of actual diffuser units needed, divide the total square foot required by the area size of the selected diffuser.

Where you place these air make-up units is also quite important. Improper placement can cause unacceptable air patterns. Don't put them too close to the hood face either on the side or in the front. A good design might have them down the center of the room with the hoods on both walls. In the case where the hoods are on one end of the room, the diffusers could

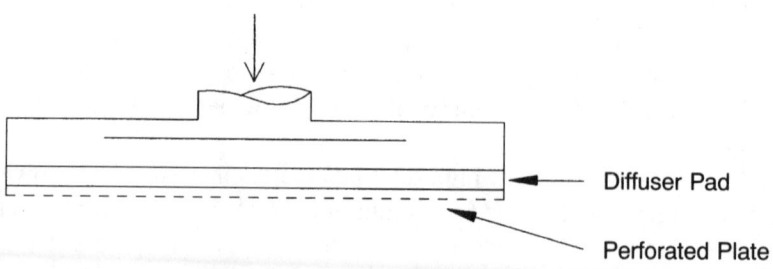

Diffuser Pad

Perforated Plate

FIGURE 2.2 *Low velocity diffuser.*

be on the opposite end. If you do not pay close attention to the type and location of the air entrance devices you will pay for it with poor hood performance. Make the HVAC engineer and yourself work to achieve the highest level of performance you can have for the hoods. That is what you are all getting paid for.

HOOD LOCATION

Now where will the ventilation and architectural people locate this big box of a fume hood? In a single floor laboratory or a multistory building with an intermediate service floor (interstitial space) they will put them wherever you want. But in a multistory building without the interstitial space the design people will want to place them next to the front doorway or a side door opening to a service corridor since this is the closest route to their vertical duct risers. This saves in the running of horizontal ducting.

If these same people were designing a kitchen in their own home (let us equate the fume hood to a kitchen stove hood), you can bet your bottom dollar that the cook of the house would not allow the stove to be in a corner or in the doorway traffic pattern. The architect, HVAC engineer, and the laboratory staff must work as a team. The hood should be located for the maximum working efficiency of the laboratory and then the other room considerations, including duct locations, will fall into their proper place.

You have spent considerable effort to slow down the make-up air. Now let us find out how much the foot traffic and door swings contribute to disrupting room air patterns. The door should always open out from the laboratory. This door pulls a large volume of air from the laboratory space and has an effective velocity of from 2 to 5 miles per hour (175 to 450 fpm). A person walking, not strolling, is traveling in the 3 to 5 mile per hour range (260 to 450 fpm). A person walking both pushes and pulls a significant volume of air. The moral is: keep your hood locations away from heavy traffic patterns.

It is not the intent of this book to design the layout of laboratories since the project people are much more knowledgeable as to equipment requirements, space allocations, and facility mission. All I am trying to do is to furnish you with the most important criteria to consider in planning for efficient and safe fume hood performance.

REFERENCE

1. *Industrial Ventilitation Manual*, 17th edition, American Conference of Governmental Industrial Hygienists, Cincinnati, 1982.

3

Basic Hood Design

If you were going to design and build a successful automobile you would first establish all the requirements: safety, economy, reliability, maintenance, load capacity, cost and so on. There is no use in specifying a 350-hp engine and putting in a midget drive train; the wrong alloy in the frame and it may crack; a plastic body that may warp or spark plugs that cannot be replaced.

How do you avoid these problems? You do so by understanding what each component does and then start making judgment calls as to what you want and how you are going to achieve it. We are not designing or building cars, there are enough people in that market. In this book we are concerned about building fume hoods. To do so we must set up a checklist of our requirements and be sure that what we want is what is built. Anything less than the safest and best is just not acceptable.

We reasonably understand automobiles because we are constantly made aware of what makes them tick. However, in the world of fume hoods we are not quite so fortunate. There are some good published fume hood articles and papers, and there are some very poor ones. How do you judge the good and the bad? Now we will start our bare bones analysis of the fume hood and see how all the various parts really fit together. You will then be in a better position to make the judgment calls on the various articles yourself and with a pretty good level of confidence.

For purposes of clarification, and to be sure we are all calling the same part by the same name, let's review accepted fume hood nomenclature as shown in Figure 3.1.

SIZE

The size of a hood, say, 6 ft, refers to the outside (side to side) dimension of 6 ft, or very close to it. The interior work chamber will be 6 ft less the thicknesses of the two side walls, which range from $3\frac{1}{2}$ to 6 in. per side. The depth of the hood again is measured from the outside shell and can vary from roughly 32 to 37 in. depending on the manufacturer and on the particular model or type. The depth of the interior work chamber is primarily affected by the air foil design and back baffle design and is partially influenced by the laboratory workbench depth.

Why worry about size? First it is easy to make the mistake that a 6 ft hood has 6 ft of interior work space. Secondly, the architect/engineer many times selects the size based on allocated floor space. To me that is the tail wagging the dog. The laboratory must have hoods compatible with user

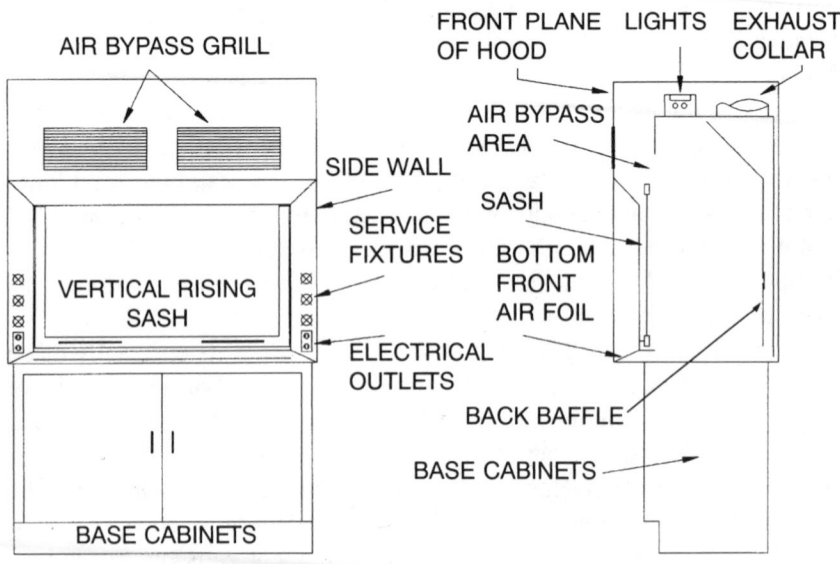

FIGURE 3.1 *Fume hood nomenclature.*

requirements. If a 5 ft hood is sufficient, that's great. However, if you really need an 8 ft hood for your equipment and you wind up with a 5 ft hood, you have a real problem. Tell the planners what you need and somehow they will figure out how to provide it for you. I am firmly convinced that if you do a really good job of educating the architect/engineer as to your true and viable requirements, you will get what you need. Try a snow job and it always comes back to haunt you.

The more basic questions we should be asking, however, are not those of overall size but what is required to provide maximum performance: then see how much space it consumes.

SIDE WALL CONSTRUCTION

Let us consider the hood side walls. They serve two purposes: (1) to provide ample space for the mechanical and electrical services (i.e., air, gas, vacuum, 120 volts, etc.) and (2) to provide the dimensional detail for the aerodynamically shaped entrance to the hood chamber.

In considering the performance improvement that is supplied to the hood due to the side wall, smaller is not better. Some manufacturers can provide as small as a $3\frac{1}{2}$ in. thickness; however, most prefer to furnish a $4\frac{1}{2}$ to 6 in. thickness. The plumber really appreciates as much space as he or she can get so as to run the piping and your maintenance people will bless you for a larger size should repairs be required. All that aside, the real reason to consider a maximum thickness is that the thicker the wall, the more gentle the continuous slope of the aerodynamically shaped entrance. The more gentle the slope the less turbulence is caused by the vertical front edges of the opening. The less turbulence, the better the hood performance, refer to Figure 3.2.

My recommendation is for a $4\frac{1}{2}$ to 5 in. dimension with a continuous slope. This has a noticeable effect on improving hood performance.

HOOD DEPTH

While the side-to-side dimension of the hood is critical for equipment placement and use, the front-to-back size is more critical and less forgiving from a safety standpoint.

Depending on the manufacturer this overall dimension ranges from 32

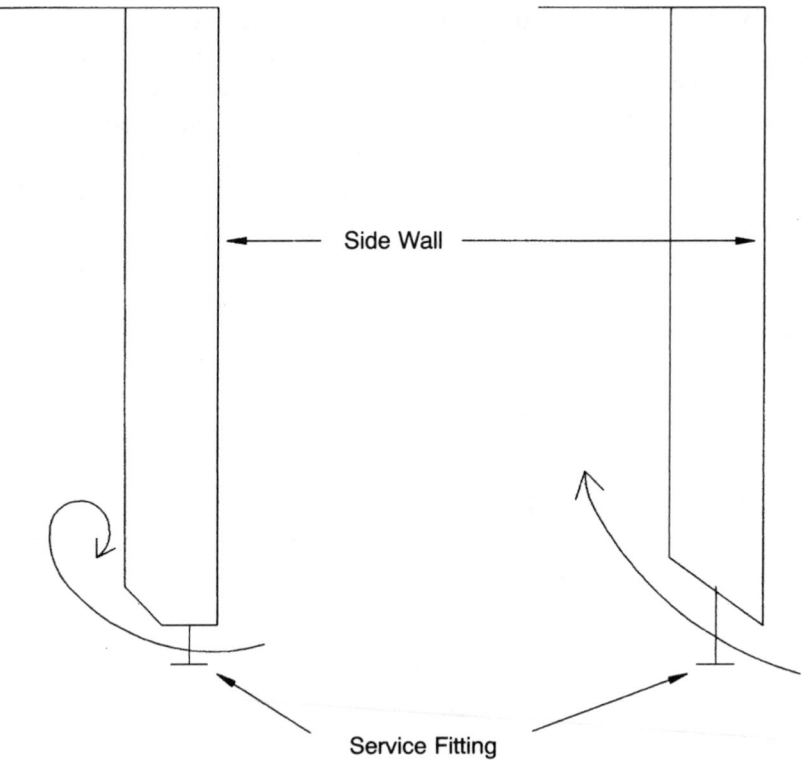

Side Wall

Service Fitting

FIGURE 3.2 *Hood side wall design.*

to 37 in. Architects also have some influence here when they do the laboratory layout. Depending on workbench depth and aisle space there may be a demand to keep the hood depth as small as possible. This is not an area that should be compromised. First, you jeopardize safety and then you get caught up with too little space in which to work. So let's design the depth and include all those critical items that make up this front-to-back profile. The six segments of this part of the fume hood are outlined below and shown in Figure 3.3.

Quickly, let us consider our working surface allowance. If the hood has an overall depth of 37 in. the safe usable work space will be approximately 21 in. Dimensions A, C, and D (Figure 3.3) are most critical for superior hood performance. Each of these design parameters will be explored in detail later in this chapter.

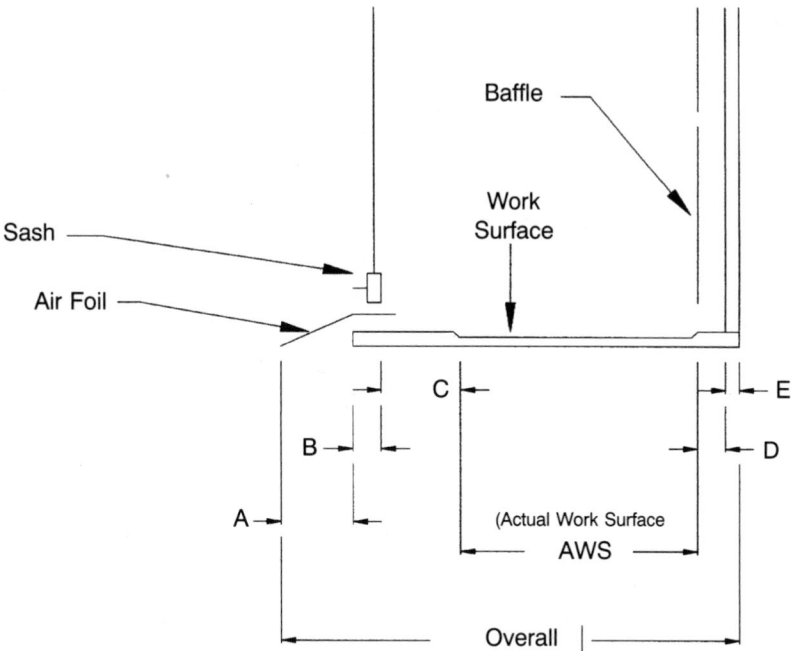

FIGURE 3.3   *Six critical design points for fume hood. (A) Air foil depth ($4\frac{1}{2}$ to 5 in.).
(B) Sash thickness and set back from front edge of hood chamber, depends on both
the sash and sash track construction (1 to $1\frac{1}{2}$ in.). (C) Recommended minimum
work-free zone behind the sash for maximum safety (6 in.). (D) Air (duct) space
between the back of the baffle and the hood lining (minimum of 2 in.). (E) Allowance
for framing system, material thicknesses, and wall clearance 1 to $1\frac{1}{2}$ in.*

INTERIOR CHAMBER DIMENSIONS

The interior vertical size can be about any height that your procedures
dictate. The standard manufactured dimension is approximately 48 in.;
past history has sustained this height as being adequate for most operations. Manufacturers like it because it conforms to the 48-in. width of the
uncut sheets of most hood lining materials. Their savings in materials are
reflected in their pricing. Add any amount of height and the cost increase
may seem disproportionately high. You must remember that the manufacturer has to make a special liner, an entirely special outside shell and
baffle system, and maybe even a special sash. If you plan to have the
interior height larger than 60 in. you should consider walk-in or distillation

hood designs, since your back baffle system could become self-defeating due to improper slot design (refer to pages 17 to 23 and 41).

SASH OPENING SIZE

What other factors should we consider in discussing the height of the fume hood superstructure? For starters how about clear sash opening. Hood fabricators make their products in a manner most convenient to their manufacturing processes and tooling. By reviewing Figure 3.4 we shall arrive at some viable user requirements.

If: A = 72 in.

then: B = approximately 5 in. less than A.
       C = 29 to 30 in.
       D = 37 to 38 in.

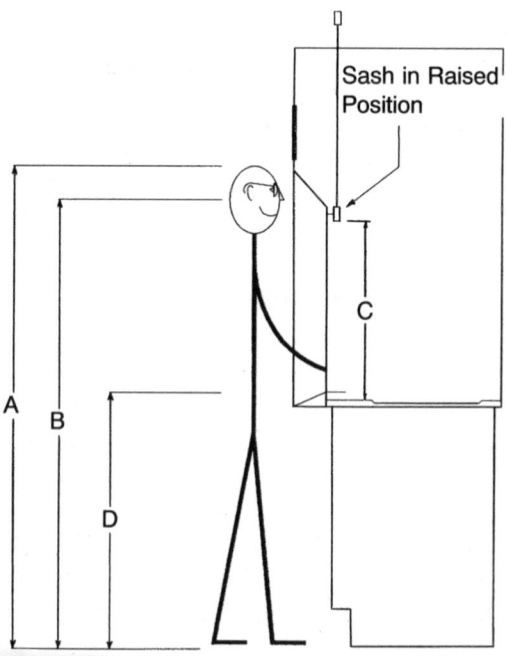

FIGURE 3.4   *User requirements regarding sash opening sizing.*

Dimension C is the opening of the sash. It should be at least 29 to 30 in. If you compare this opening size in various hood catalogs you will see that the clear opening size varies from 26 to 31 in. People are getting taller with each passing year, and I feel you should insist on at least a 31 in. clear opening; I have seen some rather large facilities insist on 33 in. If you have a large quantity of hoods the price difference between 31 and 33 in. is small. The added inch or two does not adversely affect the operational safety of the hood proper.

We have considered macro design; now let us down-size to our individual hood components.

AIR FOIL DESIGN

A fume hood that does not incorporate the air foil entrance shape is roughly 50 years behind the times. A square faced hood is an accident that is happening all the time even under pretty ideal conditions. Given a laboratory with some noticeable room air movement or personnel traffic problems, you can have a real disaster safety-wise when hoods lack the air foil design.

The most important section of the aerodynamic entrance shape is the bottom front air foil. The reason is very simple and easy to demonstrate (Figure 3.5).

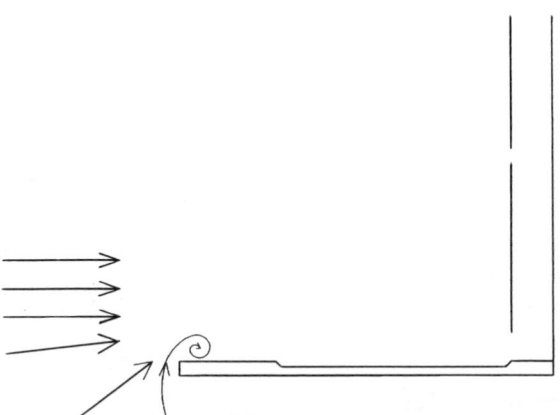

FIGURE 3.5   *Front edge turbulance without bottom front air foil.*

Remember one of the statements in the introduction was that "air is lazy." As air goes into a hood it arrives with both a horizontal and a vertical vector component, and these collide at the entrance to the hood at the work surface area. This forms a large roll that sits at the front edge of the hood and can (and does) bring hood contaminants from the interior of the hood to the front. If someone walks past the hood or there is a room cross-draft, your hood-contaminated air is being mixed with the room air.

This is easy to demonstrate. Take a pie pan and fill it with $\frac{1}{2}$ in. of warm water and place it about 3" back from the face of the hood. Break up some dry ice and put several reasonably small pieces in the water and see what happens to the CO_2 and water vapor [1]. A hood with no air foil has the turbulence well defined. Add an air foil of the proper design and "poof" the turbulent pattern just goes away (Figure 3.6). As repeated in many advertisements, "try it, you'll like it" or "don't leave home without one."

Now let us explore air foil design. Are they all created equal? Not really.

The slope of the foil should be reasonably gentle, say, an angle of 20° (Figure 3.7). Thirty degrees is no disaster, but you can get into trouble at 45°, especially if the spacing under the foil is greater than the normal 1 in.

Dimension A is pretty standard at a nominal 1 in. (1 in. less the metal thickness). Individual laboratory requirements due to electric plugs, tubing, and so on, where you want these to exit the hood under the foil, may require a spacing of $1\frac{1}{4}$ in. This is no real problem as long as dimension B is large enough to straighten out the air flow. When the space exceeds $1\frac{1}{4}$ in., you enter a problem area and could have a roll of turbulent air at the front edge under the foil.

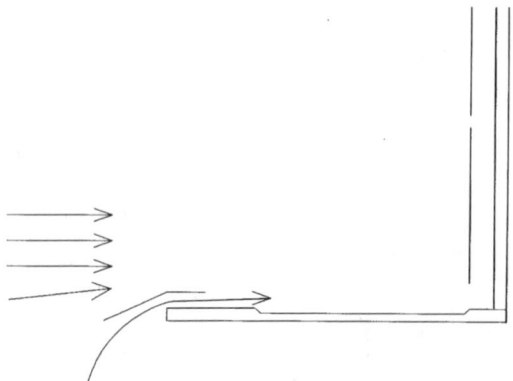

FIGURE 3.6 *No turbulance with bottom front air foil.*

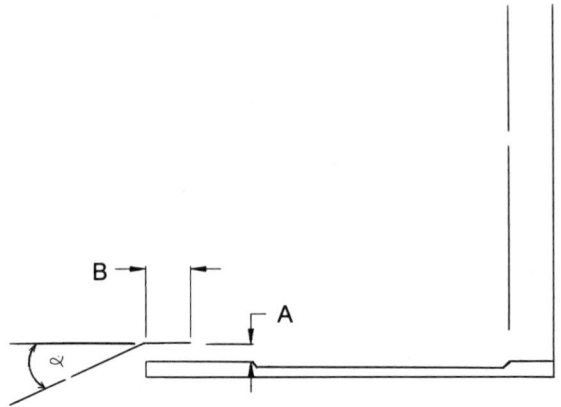

FIGURE 3.7   Air foil design.

Manufacturers do some silly things with this vital design parameter. One makes B too small to be really effective; another makes A too large and B too small and for rigidity puts a downward slanted return on the back edge of the foil, making it worthless for all practical purposes.

You should also consider the materials of construction of this front foil. One that is made of mild (regular) steel and painted with the same finish as the laboratory cabinets is a mess in a year or less. A stainless steel foil used in an acid environment takes a good beating because stainless steel just isn't stainless. While stainless steel can get pretty grubby after awhile, you can clean it up with some cleanser and elbow grease. By far the best is one made from mild steel and covered with a special coating that is specifically chosen for the procedures used in your laboratory. My experience tends to favor Kynar [2], however, pinhole-free Teflon [3] would be a very good choice. Teflon can be white where Kynar is a dark grey.

BACK BAFFLES

Referring back to the introduction, we said that the back baffle was the first major improvement to the alchemist's fireplace. Not only was it the first, but it may well prove to be the most important part of an operating hood (Figure 3.8).

As hoods are currently manufactured there are two slots (A and C) or three slots (A, B, and C) in the back baffle. A and C are normally adjustable

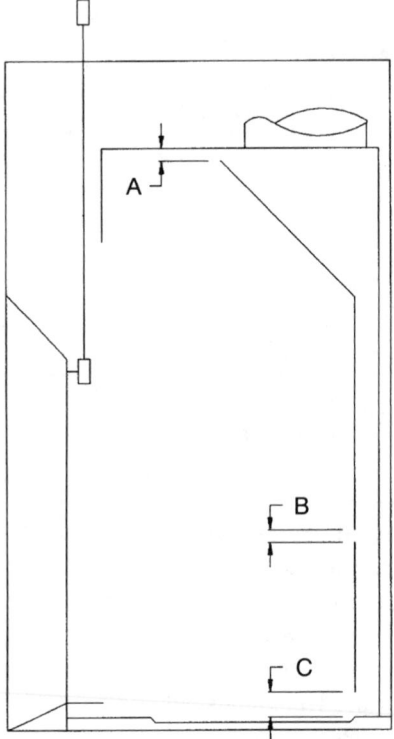

FIGURE 3.8 *Standard back baffle slot sizing.*

from a closed position to full open (roughly 2 in.). Slot B is usually a fixed size and varies from 1 to 1½in. Product catalogs state that these slots are adjustable to compensate for heavier than air and lighter-than-air gases. So much bunkum. Except for heat and natural gas it is rare indeed to utilize a lighter-than-air reactive gas. Most vapors, including solvents, are heavier-than-air. Procedures that utilize mercury or bromine produce a substantially heavier than air environment. Admittedly you have a great deal of dilution of reaction chemicals and products. But even so the average off-gas is slightly heavier than air even in a dilute state. Nonheated reactions are preordained to be more dense than their heated brethren.

Since we are primarily dealing with heavier than air products how do we adjust these "adjustable" baffles? Short answer: take the bottom "adjustable" part off and throw it away. The top opening should be set in the

$\frac{1}{2}$-in. range and then screwed or glued in place so that it cannot be changed by an uninformed worker.

Not convinced? Okay, then let's see how a fume hood works in regard to various baffle slot settings (Figure 3.9) [4], [5].

Your principal exhaust action is past and over your work surface and the balance goes through the middle slot B and top slot A. With A open to $\frac{1}{2}$ in. you generate an area of swirl called the vortex. At this $\frac{1}{2}$ in. setting the vortex is confined to that volume in the top portion of the hood chamber that is behind and level with the bottom of the raised sash. Your work surface is well ventilated; we refer to this as "floor sweep" (Figure 3.10).

The top slot A receives the maximum exhaust potential since it is closest to the exhaust duct. When open it really diminishes the volume of air available to B, but most especially to C. The floor sweep that is essential

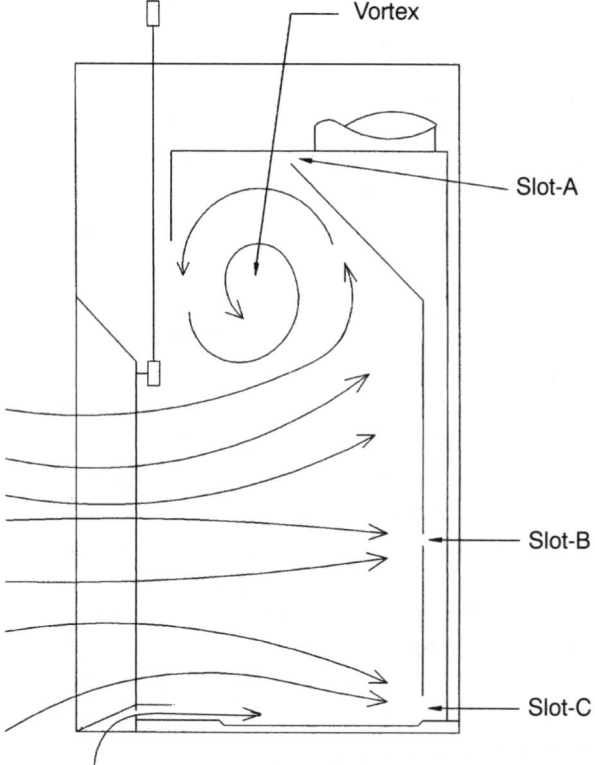

FIGURE 3.9 *Baffle slot settings. Conditions: A open $\frac{1}{2}$ in.; C wide open (2 in.).*

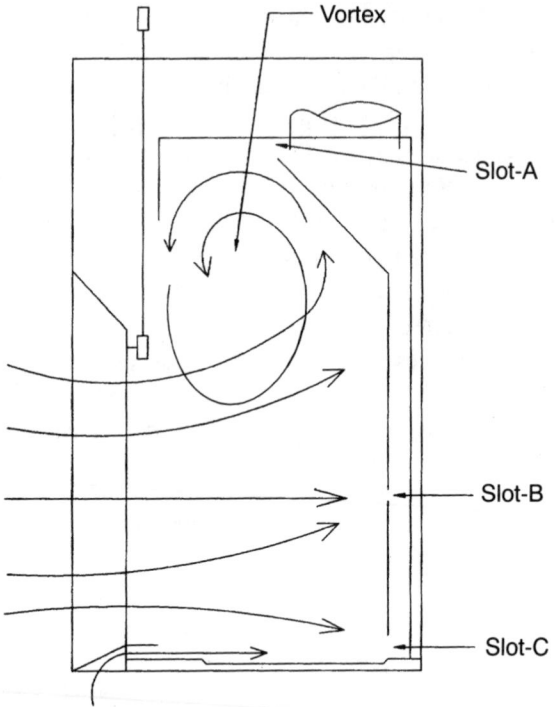

FIGURE 3.10   *Ventilation at work surface. Conditions: A wide open (2 in.); C wide open (2 in.).*

for safe fume hood operation is now marginal or worse. You concurrently have a substantial increase in the size and the amount of turbulence in the vortex area. Why is this? Well, the air that was hoping to be exhausted at A has acquired a fairly high velocity and speeds right past slot A. Due to the hood geometry it heads back down into the hood chamber where it is buffeted by the horizontally directed air just entering the front face of the hood at the bottom of the raised sash. hence the large swirl. With A wide open the vortex extends down below the bottom of the open sash, where it is subjected to more abuse.

The cross-drafts that are present in a laboratory regardless of the source can and do grab some of this interior hood air and force it into the laboratory proper.

The area located by the bottom of the raised sash is basically that of a square corner. This area also pulls air forward, and unfortunately this is about the right spot for a person's breathing zone. For this condition it is easy to see that the hood has become an unsafe area in which to work. It can, however, get worse, as we will see in Figure 3.11.

You now have no floor sweep in the hood. The size and degree of the turbulence of the vortex are now affecting the hood air pattern well down into the hood chamber. Any room disturbances, even those that would normally be acceptable for safe hood operation, totally disrupt the air exhaust pattern and allow copious quantities of hood-contaminated air to enter the laboratory.

What does this all add up to design-wise for the back baffle?

1. We do away with the adjustable slot closures at A and C.
2. We design A to be $\frac{1}{2}$ in.

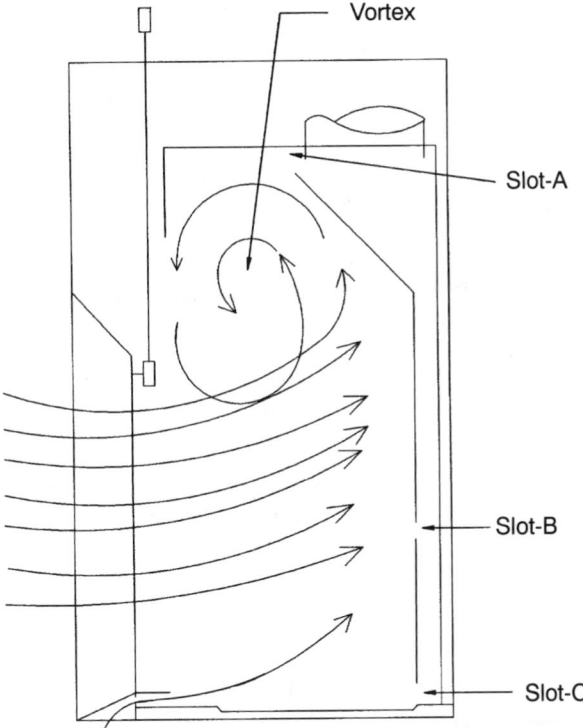

FIGURE 3.11 *Unsafe hood conditions: A wide open (2 in.); C closed.*

3. Slot B is fixed at $1\frac{1}{2}$ in. and located 14 in. above the work surface.

4. Slot C is a fixed slot 2 to $2\frac{1}{2}$ in. wide.

5. The clear air space behind the back baffle is 2 in.

While we are discussing the back baffle let us now consider another important design parameter (Figure 3.12).

Once slot A is set the body of the back baffle should be forward of the exhaust duct opening by at least 1 in., the reason being that the exhaust duct area has the greatest exhaust suction that is available to the hood.

If dimension X does not exist (no overlap), you have a much larger volume of air being exhausted in the center one-third of the hood as compared to the side wall areas. This cuts down on side wall ventilation and allows more turbulence to occur at these vertical leading edges.

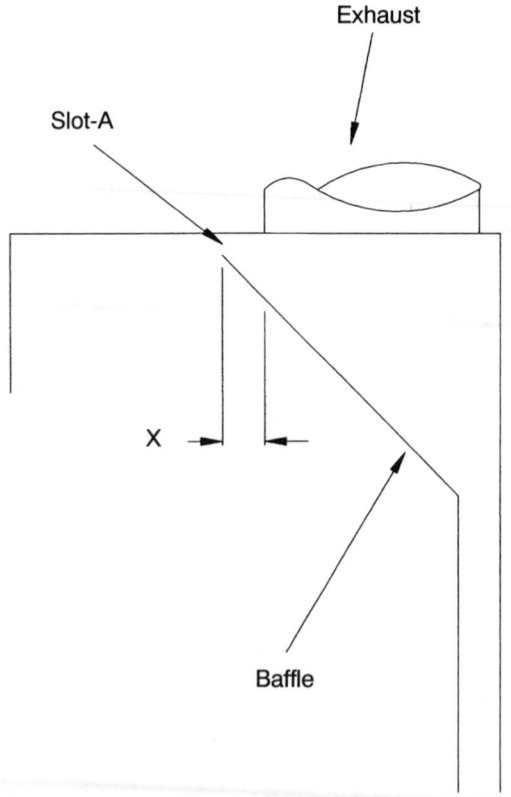

FIGURE 3.12 *Back baffle plenum design.*

In Chapter 9 we will be reviewing a large number of field tests of hoods (ASHRAE protocol). One of the first examples is the effect of slot openings on performance. If at this time the ASHRAE protocol is a "puzzlement" (àla "King and I"), be patient as we will also do a bare bones review of this important procedure.

We have reviewed the two major design areas of the hood superstructure: the air foil and back baffle. Now we move on to the sash and sash area.

VERTICAL SASH

The front sash offers an extension of safety that cannot be found on a full open-faced hood. If things start to go wrong and you need a quick place to hide, then the lowered sash offers this haven. It has other attributes, but for now we are going to be very general.

The sash is normally framed and fits into tracks in the hood side wall. It is counterbalanced to be reasonably easy to raise and lower and the glazing material is normally $\frac{1}{4}$ in. safety plate. The frame material can be stainless or painted steel. It makes little difference and depends on the user's preference and aesthetic requirements. The interior, or retaining part of the assembly, is another story. It should be a coated steel (Kynar or equivalent) and not painted or stainless steel. Rigid plastics compatible to your procedures are also very acceptable.

The two most annoying problems with a sash are (1) it is hard to move up and down and (2) it binds in the sash track.

The sash should be easy to move if the sash weights are of the proper size, and the pulley system is uncomplicated and well lubricated. You may not get, or need, ease of operation with a one finger push, sashes all have handles so you really are looking at a hand push/pull, not one-finger. It is more important to have the sash come down easily. If you have it set at a mid position, it should not creep either up or down.

The sash assembly fits into the track system with some type of spacer system. A well designed system is almost a guarantee of good operation. They should be simple and not require some type of field adjustments. If your sash binds due to the spacer/track alignment, it may not be the design but a poor installation of the hood in the laboratory. This is easy to check. With a measuring tape see if the distance between the side walls is the same at the top of the sash area as it is at the air foil area (plus or minus $\frac{1}{16}$ in.). If the hood opening is canted in or out down by the air foil, it isn't too hard to fix. Side walls will move a little if you persuade them with a hammer and a block of wood. You can most likely move them up to $\frac{1}{4}$ in. without causing any structural problems to the hood or the service piping. If you can't correct the problem, call the manufacturer. They would much prefer to have you as a friend than as a complaining user. But for heaven's sake call them when you first have the problem, not 10 years later.

HORIZONTAL SASH

A horizontal sash only moves from side to side, it has no vertical motion. The glass panels may or may not be framed. They are normally hung from a roller and track system and each panel overlaps its neighbors by at least $\frac{1}{2}$ in.

Horizontal sashes have suffered from poor reviews since they were first introduced and this has to be at least 75 years ago. The primary reason was that the original design was grossly deficient. The number of horizontal panels that were used were just too few. Let us use some examples to show what happened (Figures 3.13 to 3.15).

There is no sense in drawing any more sash configurations as it is now obvious that the number of sashes and tracks can make or break a horizontal sash system. You can have five sashes in three tracks, eight sashes

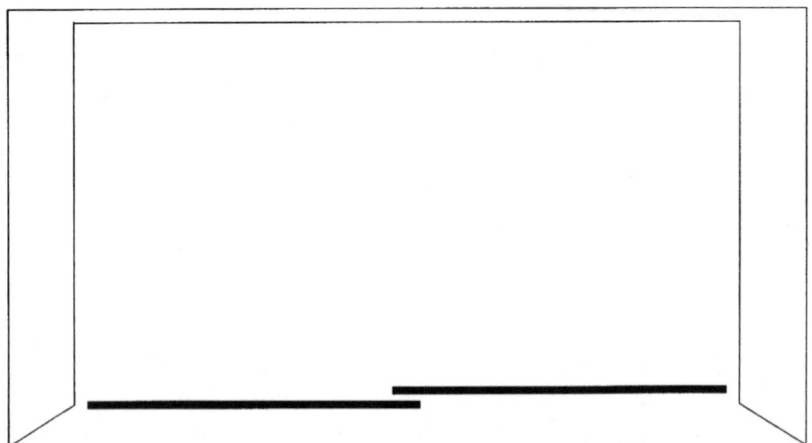

FIGURE 3.13 *Two sashes, two tracks. (1) You cannot reach the center area of the hood. (2) This barely provides a 50 percent maximum open sash area.*

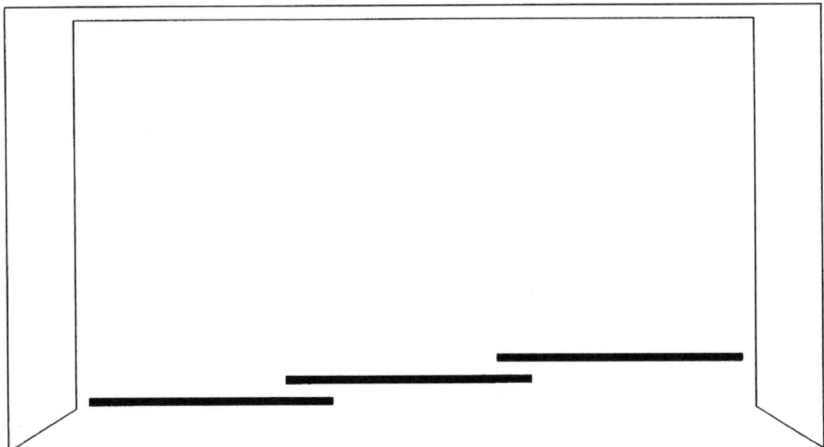

FIGURE 3.14 *Three sashes, three tracks. (1) You can reach the center area of the hood. (2) This provides 66 percent maximum open sash area.*

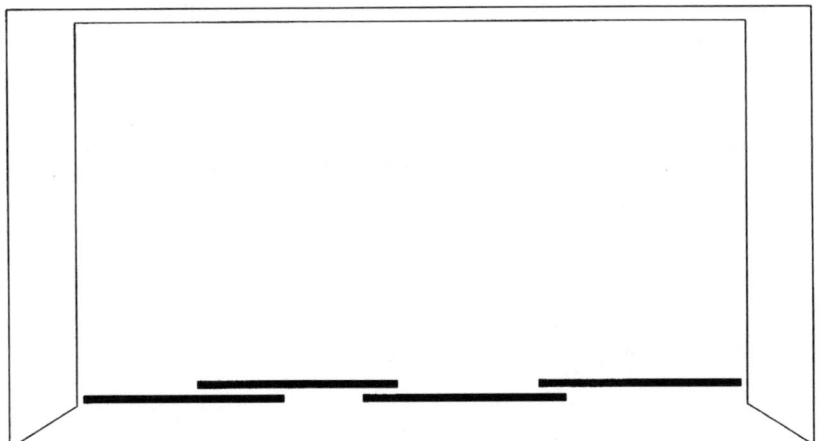

FIGURE 3.15 *Four sashes, two tracks. (1) You can reach the center area of the hood. (2) Panels could be made small enough (15 in.) to act as safety shields. (3) This provides 50 percent maximum open sash area.*

in three tracks, or about anything you want depending on the width of the hood. Never make the sashes smaller than 14 or 15 in. as they just don't slide well. However, the 14- or 15-in. design allows the sashes to act as a series of movable safety shields. Over 15 in. is not comfortable for most people to reach around with both arms simultaneously.

A hood smaller than 6 ft should not be considered for a horizontal sash design. The opening is too small. Any size 6 ft and larger can work great; however, you must insist that the manufacturer provide a very strong track support and bracing system or you can have a lot of sag in the center, and the system will not work.

The primary reason for the use of the horizontal sash system is to effect a savings of the high-quality air (heated, cooled, and filtered) that you supply to the room and then exhaust through the hood. This air is very energy intensive and saving a substantial quantity of air is saving a lot of dollars. Not only can you save on laboratory operating costs on a continuing basis, but it also saves many dollars up front in reduced capital cost for mechanical equipment and it allows for smaller duct sizes. Each reason is good news to the mechanical engineer and to the vice president of finance.

To complete our evaluation of the horizontal versus vertical sash let us use our stick figures to do the talking (Figures 3.16 and 3.17).

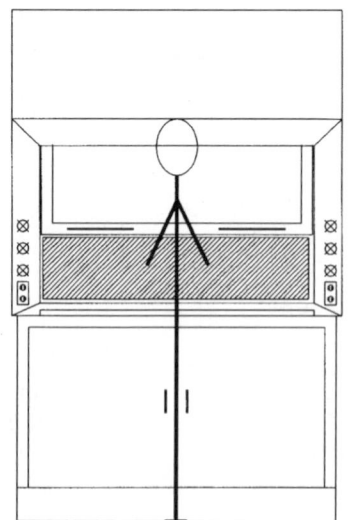 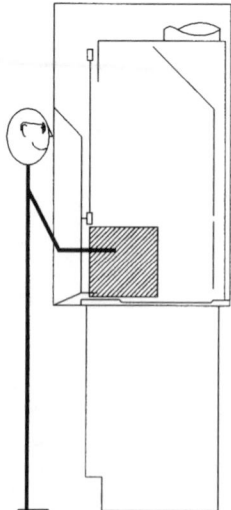

FIGURE 3.16 *Horizontal sash, 50 percent open. (1) Horizontally oriented access to hood. (2) Activity confined to cross-hatched area.*

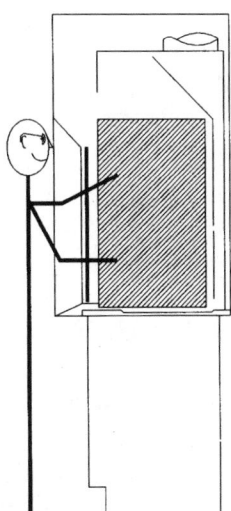

FIGURE 3.17 *Horizontal sash, 50 percent open. (1) Vertically oriented access to hood. (2) Activity confined to cross-hatched area.*

If your laboratory mission does not require any vertical reach and involves small work-top procedures, then a horizontal sash system could be an annoyance. If, however, you do need the vertical reach and you do want the savings incurred with a 50 percent reduction in air volume, then a horizontal sash system might just be for you.

If I am asking a laboratory complex to consider the horizontal sash system, I suggest that they have their maintenance shop make a mock-up for evaluation by their laboratory staff. They can use wood, aluminum channel, and plastic panels and make it so it fits, and is clamped, into the face opening of an existing hood. You will be amazed at how many converts, even supporters, you can make.

HORIZONTAL/VERTICAL

The combination of a horizontal system in a vertical rising sash is a hybrid that seems to be the answer to a laboratory group with multifaceted hood requirements.

In this system the vertical sash frame has the horizontal panel system built into the framing. They are normally unframed panels and are bottom supported. This costs a bit more for the hoods than the single purpose systems, but the payback is great and is a wise investment for diverse and changing laboratory assignments. The Monsanto Research Laboratories in St. Louis have installed this design, to the exclusion of all others, for the past 10 to 15 years.

SASH AREA

I use this connotation so we can explore other areas in or around the sash proper.

Where a vertical rising sash goes past the superstructure body, there is a space or gap built into the hood body so that the sash frame does not bang the hood lining every time it is raised (Figure 3.18). When the sash is in the up position, air enters the hood through this gap and is in turn ex-

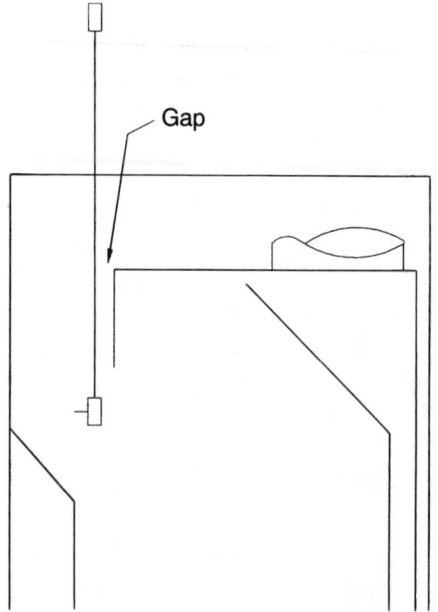

FIGURE 3.18   *Gap between hood body and sash.*

hausted through the duct system. This leakage amounts to roughly 10 percent of the total exhaust volume. If you want to save energy or increase a low face velocity (less than 60 fpm), then close this gap. It is easy to do this when the hood is built or even later in the field (Figure 3.19).

Purchase a sufficient length of 3- to 5-mil Teflon strip about 2 in. wide and fasten it to the top of the hood at the sash opening. The Teflon should ride on the glass by about $\frac{1}{8}$ in. to give a fair seal.

SASH STOPS

When using vertical sashes in an exhaust system where the mechanical engineer has only provided sufficient volume for a 50 percent sash open area, you might like to consider some sash stops. Whereas the design people depend on staff training to keep the sash at a maximum of half-open, we know that this discipline can disappear after a period of time. The sash stop is simple; it only stops the sash at a predetermined point when the sash is raised, not lowered. These stops are reminders of the half-closed (or open) philosophy, are not expensive and can be added to almost any framed sash hood at any time. They are better than nothing, and in many installations have proved quite successful (Figure 3.20).

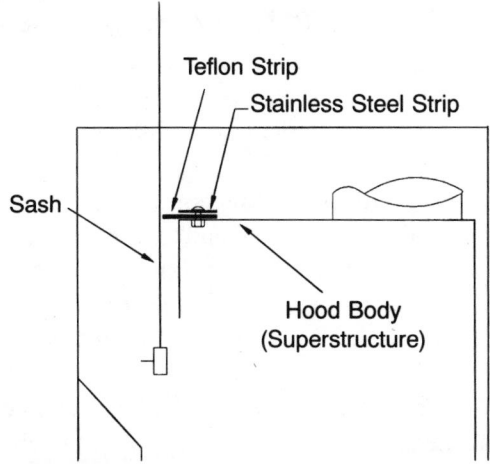

FIGURE 3.19 *Closing gap between hood-body and sash.*

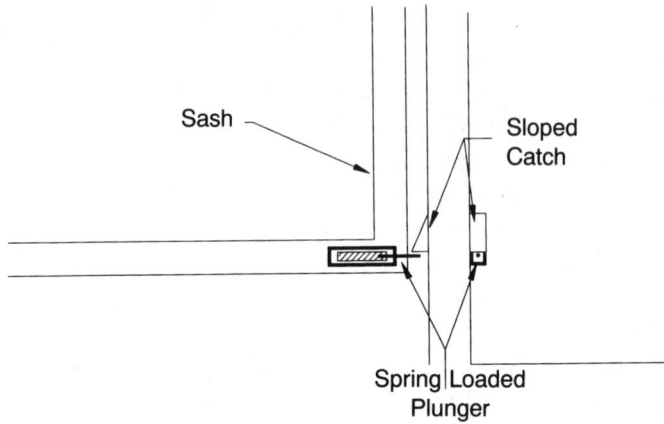

FIGURE 3.20 *Sash stop.*

BYPASS DESIGN

Horizontal and horizontal/vertical sash systems require some modifications to the bypass above the sash which is normally an unobstructed area. This type of bypass is used on all constant-volume fume hoods. When the sash is in the down position, the open area allows room air to enter the hood and maintains a reasonably constant exhaust volume.

Constant-volume hoods are NOT constant face velocity hoods; as the sash is raised from the closed position the velocity increases by a factor of from 3 to 4 depending on the clear open size of the bypass itself. The only so-called constant-velocity hoods are those equipped with variable air volume (VAV) control systems. Variable air volume systems are reviewed in Chapter 6.

Envision a horizontal/vertical combination sash with a conventional (full open) bypass area. Open one panel and then raise the sash. A large amount of air rushes into the hood by this path and not through the working sash area; the face velocity takes a real tumble. Again, you can rely on discipline to be sure all panels are closed when the sash is raised, but people can be forgetful.

This is easy to fix. Cover the bypass area with a perforated plate. Use the same material as the hood liner and with a drill or punch provide a 12 percent to 15 percent open area pattern. Why bother? If the sash is down and all the panels are closed, you still need some air ventilating the hood.

That area under the air foil is too small to provide sufficient air at an acceptable velocity.

For a horizontal sash system the bypass area must also be fitted with a closure panel. Use the same material as the liner, only this time drill or punch a series of $\frac{1}{8}$-in. holes on 1-in. centers across a 2-in. central horizontal area of the closure panel. This too serves to give you upper hood ventilation when all the sash panels are closed.

MECHANICAL SERVICES

We will not delve into a lot of valve detail; several companies describe these products quite well in their catalogs. The hood manufacturers usually include only one brand of fixtures in their catalogs but can supply other brands and styles as made by any of the major fixture houses. Have your mechanical engineering people select the best valving system not only for you, but for the total laboratory complex. In an integrated facility it makes a lot of maintenance sense to settle on one valving system. Our aim is to use these systems to our utmost advantage, regardless of the brand.

The enclosed space in the hood side walls gave the mechanical people a chance to hide a lot of piping and wiring. It also provides a much better location, from the user's standpoint, to mount the service valves. The valve handles are up where you can see and reach them easily.

In the last few years some manufacturers have been producing hood valves where the valve body is mounted on the front surface of the vertical facia panel rather than being located in the enclosed space and being operated by a remote control rod. These are referred to as front-mounted valves; however, it is not a vital consideration in regard to hood performance.

Figure 3.21A shows a typical remote control rod valve assembly. If the valve starts to leak, it is hard to fix and the user has some diminished fine control over the valve settings. Figure 3.21B shows a front-mounted valve assembly with the valve stem and handle oriented straight out from the facia panel on the hood. Figure 3.21C shows a front-mounted valve assembly where the valve stem and handle assume the angle of the facia panel proper.

I am favorably impressed with the system shown in Figure 3.21C. The valve identification button is easier to see; such valves are easier to operate and control is better. The valve handle only protrudes about a $\frac{1}{2}$ in. past the

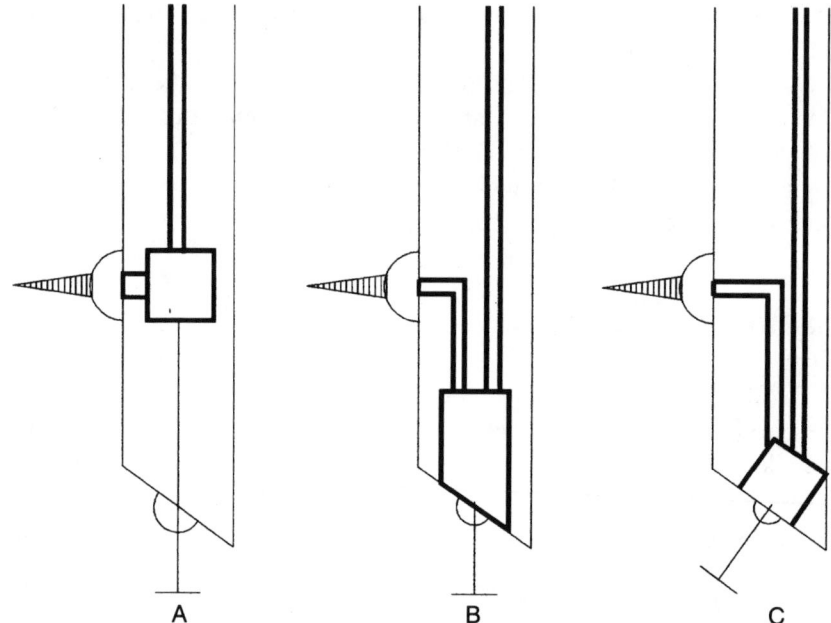

A   B   C

FIGURE 3.21   *Hood valve assemblies.*

front plane of the hood. Remote control rod valves and straight oriented valves stick out at least $2\frac{1}{2}$ in. With that much reach they are great at grabbing a lab coat.

The interior hose bibs and faucets should get some attention if only because of their finishes. Chrome-plated fixtures deteriorate with time. Those with an aluminum or bronze finish do much better and those with special epoxy finishes will out last the other two. You can get molded nylon and rigid PVC bibs and these can last a long time.

For hoods that exhaust fumes from perchloric acid digestions, commercial coatings are not compatible and chrome plating corrodes rapidly. Nylon is not acceptable; molded PVC is compatible and will last as long as the hood itself.

One last suggestion. Try to eliminate all metal that is not specially coated from the interior of general purpose hoods. This includes screws, nuts, and bolts. They rust and ruin the appearance of the hood and may present cross-contamination.

REFERENCES

1. John B. Adams, "Synthesis Laboratory Fume Hoods: Easy Reliable Performance Evaluation; Importance of Sash Design." Stine-Haskell Research Center, E.I. duPont, Newark, DE, 1988.
2. Kynar, Pennwalt Corp., 3 Parkway, Philadelphia, PA.
3. Teflon, E.I. duPont Co., Wilmington, DE.
4. G. W. Knutson, "Effect of Slot Position on Laboratory Fume Hood Performance," *Heating, Piping and Air Conditioning*, 56, 93 (February 1984).
5. G. T. Saunders, "A No-Cost Method of Improving Fume Hood Performance," *Am. Lab.*, 102 (June 1984).

4

Applied Product Design

In Chapter 3, Basic Hood Design, we defined the various parameters that govern good overall hood design and performance. While these criteria are dominant for all fume hoods, there are specifics of construction that are applicable to special purpose hoods. It is this area of "specialty" that we now explore so as to define and construct hoods for specific applications. We also review some of the "appliances" that can be, and in some cases must be, part and parcel of most hoods.

PERCHLORIC ACID HOODS

The perchloric acid hood is a single usage hood. It should only be used for perchloric acid digestions, and under no circumstances should it be considered for general purpose chemical procedures. Perchlorates formed during the digestion process are very unstable and when dry have a very rapid reaction rate that can cause violent explosions. Fortunately, these perchlorates are water soluble and we use this property to help contain the hazards of this chemical.

A perchloric acid fume hood, and its associated exhaust system, must meet these three basic parameters:

1. Made of materials resistant to reaction with perchloric acid.
2. Watertight construction to accommodate water wash down.
3. Void of sources of spark generation.

There are two materials that can be used for fabricating these hoods: type 316 stainless steel and rigid PVC.

Stainless steel is the material of choice by fume hood manufacturers since they have facilities to weld and polish this material. The resulting product is initially attractive and is structurally sound. Stainless steel is not really stainless, of course, when it comes to most acid applications and in a year or so the hood appearance is no longer pretty and shiny. It is still very safe to use, but pretty it is not.

Rigid PVC, on the other hand, is as attractive after a year of service as it was on day one. There is one major problem with securing hoods made of PVC; major hood manufacturers do not have the expertise or facilities to construct hoods from this material. The hood superstructure commonly is subcontracted to an outside firm with little or no knowledge of fume hoods but with expertise in plastic fabrication. Sometimes the results are marginal, sometimes great, and it all depends on the coordination between the hood manufacturer and the PVC fabricator. In some geographical areas there can be local fire codes that preclude the use of PVC, but it may be possible to secure a waiver. Have your hood supplier expend the effort to coordinate the engineering and construction with the outside fabricator, and you will be more than compensated for the effort. At the end of several years of service this material looks like new, whereas the stainless steel hood may have been replaced at least one time.

The water wash-down system, in the hood superstructure, must be designed and constructed so it provides complete coverage of the area behind the back baffle. It must do this without spraying water over the baffle system and onto the working surface of the hood. Some fabricators use spray nozzles, some perforated pipe; it has been my experience that a properly engineered and drilled pipe system can offer the best solution, (refer to Figure 4.1).

Remember that the ducting also needs an equally efficient wash-down system and here the spray nozzles do a very fine job. This ducting wash down system must include all the ducting from the hood to the blower and the exhaust stack from the blower to the environs. The blower does not need a separate wash-down nozzle because it is the recipient of the exhaust

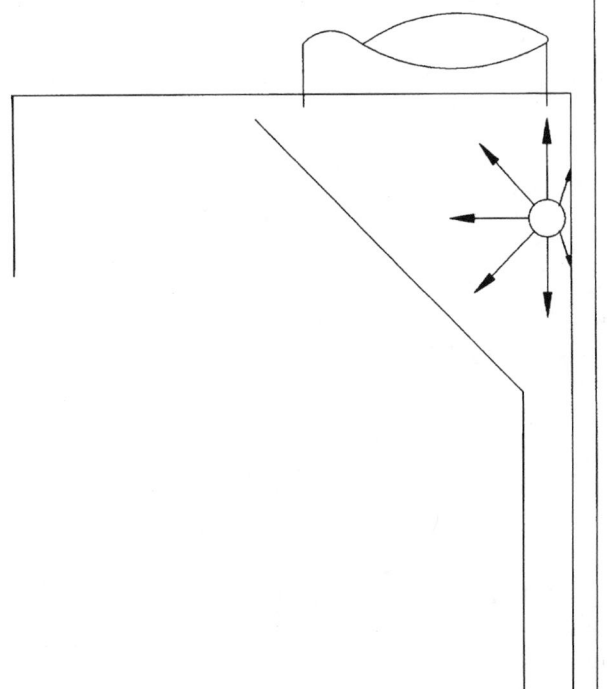

FIGURE 4.1   *Spray-down pipe design for perchloric acid hood.*

stack water. The blower needs a drain that is connected back into the ducting to the hood so that the water is properly handled.

The proper design of these hoods provides a trough along the back of the work surface to accept the wash-down waters. Just be very careful in the design stage that the mechanical engineer does not provide more wash-down water than the drain can accept. If necessary the hood manufacturer will provide a larger drain connection to accommodate the required water volume.

The lighting fixture in these hoods is normally of explosion proof, incandescent design to take advantage of the waterproof and spark-proof features; however, it does not require that the wiring to the power source meet explosion-proof codes. All electrical outlets must be external to the hood work chamber and mounted in the hood vertical facia panels or some other outside surface.

Internal plumbing fittings such as serrated hose bibs, nozzles, and the like should be molded PVC. There are coated fixtures that are used by some suppliers, but these coatings are of an organic base and provide a source of hazard. Molded nylon is also available for plumbing fittings, but it too is combustible and provides a danger spot for perchloric acid oxidation.

The gasket material used to set the glass sash panel must also be perchloric acid resistant. Rigid PVC is normally used, although I did discover that one prominent, but now defunct, hood fabricator was using a sponge material that was not resistant to perchloric acid and did not know that it was unsafe. For your own safety please check with the manufacturer for the material being used.

Indicating manometers or low-flow alarms should be part of the hood system. Hot wire devices can be used if they have glass-enclosed sensors and operate on a 12-volt supply. When inserting a pressure probe for a manometer remember to weld the small tube into the ducting and orient it downward to at least a 45° angle. The probe piping should just barely extend into the duct so as to sense only static pressure and not both static and velocity pressure. Do not use a pitot tube, since it will indicate both velocity and static pressure and the small sensing holes can plug up rather rapidly. The connection from the ducting tube to the manometer should be standard copper or stainless tubing as it may be subjected to a very small amount of perchloric acid exposure.

The exhaust collar for the hood should extend at least $\frac{1}{4}$ in. down inside the hood chamber and be seam welded to the inside surface of the chamber. Refer Figure 4.2.

This offset allows the wash down water to flow directly into the hood. If you do not insist on this slight duct extension, the wash-down water will flow out of the ducting and spread laterally across the top of the hood and eventually will drip or flow unceremoniously onto the hood work area. It is amazing how few hood manufacturers recognize this as a required construction feature.

The frequency and duration of use for the wash-down system are an area that presents a lot of questions. Here are my suggested answers:

1. If only moderate amounts of perchloric acid are used, (100 ml or less) once each day should be sufficient.

2. If the quantity is close to 1 liter or more each day, the wash down could be done two or three times each working day at some convenient pause in the procedures.

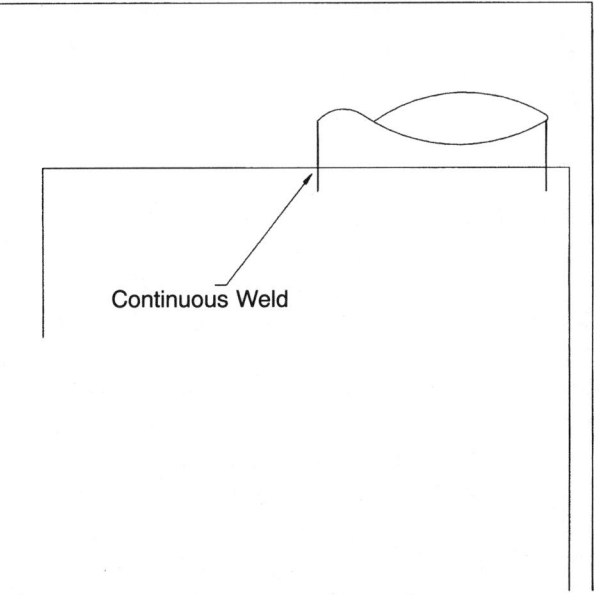

Continuous Weld

FIGURE 4.2 *Design criteria for welding and penetration of duct collar for perchloric acid hood.*

3. The wash down should be activated for a long enough period to ensure a thorough cleansing. This could be in the 5- to 15-minute range.

The wash-down system should be activated only with the exhaust blower OFF so that complete coverage can be achieved. Automated and timed wash-down systems are acceptable as long as they also turn off the blower during the water wash-down cycle. If the exhaust blower is on, you have air currents carrying the water out of the hood up the duct and away from the trough drain system built into the hood. The lack of the downflow of water also disrupts the full coverage needed to cleanse the contaminated areas.

Perchloric acid hoods do NOT fit into a central building exhaust system since they MUST have a dedicated duct and blower system. The ducting should be as short as possible and with few, if any, horizontal duct runs. Perchloric acid fume hoods are difficult, if not impossible, to locate on the lower floors of a multistory building.

I have not seen perchloric acid hoods integrated into a VAV control system. This does not mean that this is impossible; however, the numbers of this type of hood in use are quite small and the amount of air saved with VAV controls hardly seems worthy of the expenses involved.

RADIOISOTOPE FUME HOODS

The demands of the radioisotope hood have to do with two areas: cleanliness and load support for lead brick shielding. Type 304 stainless steel, cove-welded liners are most acceptable. Where the cove seams are welded they should be ground smooth and polished. The rest of the stainless liner should remain with the mill finish (referred to as a "2B finish"). The polished surfaces are shinier but much rougher than the mill finish. The work surface is normally supported by an industrial steel grating with a load-bearing capacity of 200 pounds per square foot.

Radioactive chemical hoods may require either or both particulate filters (HEPA) and activated charcoal (absorption) filters. There are reliable filter manufacturers that can supply both and will provide the proper sizes depending on the type of work being done and the volume of air to be treated. The building mechanical design engineer must be alerted to the use of these filters since their presence will require a substantial increase in available exhaust pressure for this hood system. Hood manufacturers fabricate housings for these filters; however, they are not nearly as good as those made by the filter companies. You will pay more for those supplied by the filter people but it is well worth the cost involved.

Do NOT equip a radioactive chemical hood with any type of washdown system, as it will spread any contamination into your building plumbing system. These hoods are cleaned with wipes and mild cleansers, which are then disposed of through the solid radioactive waste program.

WALK-IN FUME HOODS

Simply put, a walk-in hood is a bench hood that has been extended down to the floor level and with some minor back baffle changes. These hoods are not intended to have the user walk in and out of the hood as it is being used. They are intended to allow you to walk in, set up some equipment, and then go back outside the hood before starting any active work. They are made to house large equipment and if they are operated with the sashes closed they can work quite well.

Generally, a walk-in hood with the front sash fully open does not perform very well. They are too susceptible to room air current and laboratory traffic pattern disturbances. They should only be used for active experimentation with the sash mostly or fully closed.

The adjustment of the back baffle slots is most critical so as to ensure an adequate floor sweep and minimum vortex formation. The top baffle slot MUST be set with a maximum opening of $\frac{1}{2}$ in. and the bottom slot MUST be full open (2 in to 2 $\frac{1}{2}$ in). Anything less and you can cross off the walk-in hood as a safety device.

The sash configuration for walk-in hoods can be either double-hung vertical rising panels or horizontal sliding panels. The horizontal sash configurations are inherently safer than the vertical rising design. For starters, they are easier to move since the vertical sashes require more complex track and pulley systems, longer cables, and usually heavier weights as compared to a bench hood. Since the horizontal sashes are easier to move, you are more likely to have them closed when using the hood.

The floor design of the hood can be a critical factor in walk-in hood performance. The room air currents at floor level are much different from those at eye level. Walking, air leakage under doors, and deflected currents from make-up air diffusers can cause disturbances that are greatly amplified since the room floor and the hood entrance are pretty much at the same level. This is one reason that hood floor sweep is so important to good operation. The use of the bottom front air foil is not as important since an entrance air roll is not generated as with a bench hood.

In installations where solvents are the major ingredient in hood protocol, I have installed a barrier to help keep solvent vapors from crawling out of the hoods and onto the laboratory floor. See Figure 4.3.

The block (A) presents a physical barrier to the vapor flow from the hood and the modified air foil minimizes the effect of the turbulence caused by the block. The two components can be made as an integral one-piece assembly out of welded stainless steel, sized to fill only half of the hood opening area and hinged from the side wall area of the hood. They are raised and lowered as the use demands.

One area you should be aware of is the floor drain. It is not unusual to use fairly large volumes of solvents in walk-in hoods. If the solvent container fractures or develops a leak, you could fill your drain system

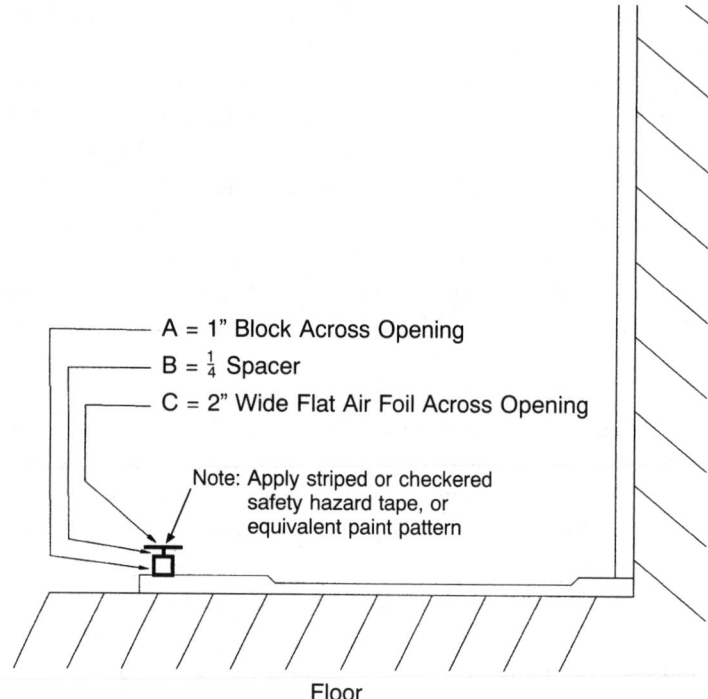

FIGURE 4.3   *Barrier control for walk-in hood.*

with a liquid that could be physically or environmentally harmful. You might have the manufacturer eliminate the drain altogether or make it so you can put in a standpipe overflow such as is used in sinks. In either case, the pan retainer design of the hood floor should be made deep enough to hold the volume of liquids that you anticipate using; the standard used by most manufacturers is for a $\frac{1}{2}$ in. indent. If you do design your pan deeper, greater than 1 in., then you probably will not need the barrier design of Figure 4.3.

CANOPY HOODS

Canopy hoods are meant to serve one primary purpose and that is to exhaust heat generated by a particular bench top group of equipment. If you have a series of ovens, it is well to exhaust this heat before it upsets your room temperature; drying ovens can also produce some unwanted odors that are not desirable to exhaust into the laboratory.

The canopy hood is not an efficient device from an air usage standpoint, but the volume of the exhaust air can be minimized. Locate the canopy hood as low as is practical for the equipment that is to be placed under it. Select a volume of exhaust air that would represent an average velocity across the opening of the canopy of from 50 to 75 fpm. This should be most adequate to remove all the generated heat and odors. Do not try to use a canopy hood as a general purpose type hood; it is not designed for this usage and will not safely contain chemical or toxic fumes.

DUCTLESS HOODS

In the past 4 or 5 years makers of the so called ductless hood have started to show a fairly substantial marketing effort. After all, a fume hood that does not require any exhaust ducting, has a built-in blower, does not upset any room air balancing, and can readily be installed by a couple of people has a lot of attractions. But like everything that is too good to be true, the ductless hood fits nicely into this category.

From a general chemical use standpoint, a hood must be able to withstand a wide variety of chemical abuses, elevated temperatures, noxious and toxic gases, and so on down a very long list of requirements. A hood system should not purposely add any type of contamination to the laboratory.

A profile of this type of hood indicates that the protection of the laboratory room from the activity in the ductless hood depends entirely on the retention capability of the particulate and absorption filters. Particulate filters (HEPA) are marvelous and have a retention efficiency for 0.3 micron (μm) sized particles of up to 99.99 percent, but they are not 100 percent efficient, and therefore some contaminants escape into the laboratory. HEPA filters do not withstand all modes of chemical attack, may not be heat resistant and can develop small leaks. The seal between the filter and the filter housing is critical for good performance. Of equal or greater importance, the activated charcoal filters that these units utilize have a minimal retention capacity at best. Absorption filters must truly be designed for the specific tasks at hand and may require flow rates and filter bed travel many times greater than are customarily available on these units.

For acceptable service the efficiency and capacity of both filter systems must constantly be monitored with a high degree of accuracy. Instrumentation to accomplish this costs many times that of this small hood itself.

The filter performance and capacity of the ductless hood are its Achilles' heel. If and when a 100 percent filter system is available then these units will become a viable laboratory hood; until then they could pose more of a hazard than a help.

GLOVED BOXES

During the 1940s and 1950s the use of localized containment systems for highly toxic and/or radioactive materials was pioneered by the University of California Radiation Laboratories in Berkeley. This rigorous containment philosophy was generally defined as concentrate, confine and control" (CCC).

The initial Berkeley gloved boxes were styled of ample size to perform a wide variety of chemical and physical procedures, but not macro chemistry involving large quantities of equipment or reagents. The chemist was isolated from the work by both the physical sides of the box and the long, almost shoulder length gloves that were sealed to the box and used to manipulate the work inside the unit. Refer to Figure 4.4.

A gloved box has its own small blower and filter system but is dis-

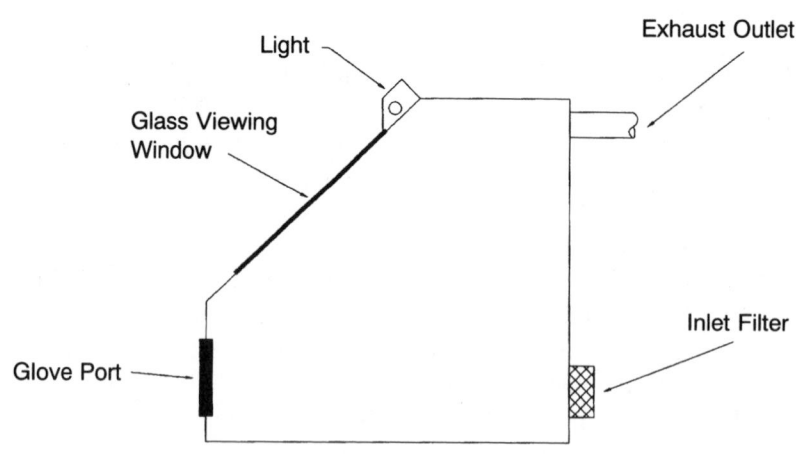

FIGURE 4.4 *Gloved box.*

charged into a central building exhaust and not back into the room. It has a slight negative pressure and the blower is in continuous operation.

While these small boxes did not lend themselves to macro manipulation, there was actually no limit as to the ultimate size as long as enough gloves were supplied to reach all of the interior equipment. I have seen some so called boxes that were all of 10 ft tall and 10 ft wide. but only four feet deep; gloved box gloves are nominally 27 in. long.

This "CCC" philosophy from Berkeley was very successfully applied to the production and/or manipulation of moisture- and oxygen-sensitive materials. There have been an untold number of very highly engineered electronic parts manufactured and special metallurgical procedures performed in this type of equipment.

The list of companies that manufacture this equipment has dwindled markedly in the past years. However, if you need help in locating a manufacturer or a systems designer refer to the *Buyers Guide* edition of some of the major trade journals.

LOCALIZED EXHAUST HOODS

Localized exhaust hoods fall into two distinct categories: One type is readily adjustable to bench top or distillation rack equipment so as to capture locally generated heat or off gases. In appearance they resemble an inverted funnel and are usually in the 10- to 12-in. diameter range. A flexible piece of tubing connects them to the building exhaust system. They are referred to by a variety of names but the most prevalent are "snorkel" or "elephant trunk." They work quite well as long as you do not expect them to do all things. Local exhausting they do, general exhaust procedures they do not do. All hood manufacturers provide these units complete with the connecting tubing, they are not expensive and are normally provided in painted steel or stainless steel.

The second type is most readily found in undergraduate chemistry laboratories. They are bench-mounted enclosures some 24 in. high and are 14 to 16 in. square, with an opening on the side that faces the student. The exhaust system is usually built up through the chemistry workbench so as to hide the ducting and give a better overall visual check for the teaching staff.

Regardless of the shape and size, they have some things in common: they really do not work well and due to their size can restrict many types of tall or wide apparatus. These undergrad hoods are mostly designed by a salesperson helped by the chemistry staff and assisted by someone with a degree in architecture. None of these people is likely to be well versed in the design of hoods and the results are unsatisfactory. When it is found that they do not provide acceptable containment, the usual response is to increase the exhaust rate, which is often more counterproductive than helpful.

This is not to say that a good localized hood for undergraduate chemistry cannot be designed, but to do so, someone will have to go back to basics, build one, test it, and go on from there. I have some suggestions where this designer can start and the direction to be taken.

Break up the turbulence that is caused by the square edges on all four sides and by the lack of a back baffle system. As a starting point I would suggest an air foil on each side, the top and the bottom. A simple back baffle system is then needed to provide both top and floor sweep in this small hood. All these suggestions are in keeping with the design parameters of Chapter 3. Refer to Figure 4.5.

I would suggest that the designers of this hood have a mock up made using any material that is handy and run some velocity and smoke tests. I feel that it will perform well with a face velocity in the 70-fpm range unless there are truly severe room cross drafts from the ventilation system or very active traffic patterns. I am certain that it will be a big improvement over what is currently being installed in undergraduate chemistry laboratories. When someone does design, test, and build a local ventilation hood that truly works, please write to me and the report will be included in a revised edition of this book 2 or 3 years down the road.

SCRUBBER SYSTEMS FOR HOODS

There are reputable manufacturers of scrubber systems for exhaust systems of laboratory fume hoods and I will not invade their design territory.

There are specific areas that could benefit from a hood scrubber and this should be determined by an industrial hygienist after a proper amount of evaluation. I will not invade their territory either.

I only want to give some first-hand advice that I have gained from observing a number of scrubber installations that had been abandoned because they were not effective with the particular chemicals being used in the hood. First and foremost the lab people often did not understand that

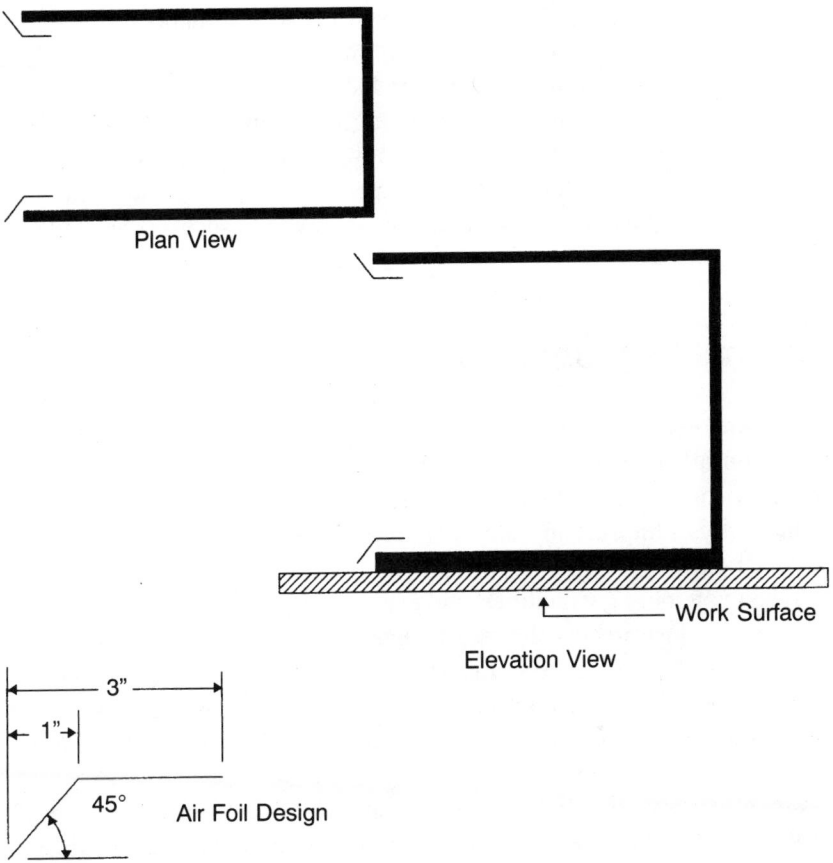

Plan View

Work Surface

Elevation View

3"

1"

45° Air Foil Design

FIGURE 4.5 *Suggestions for design of undergraduate chemistry fume hood.*

they are a piece of high-maintenance equipment and no one was prepared to service them properly.

During usage, scrubber systems need a regular clean-out schedule because they cake up and plug with the salts that form as the scrubber liquid evaporates during normal operation. This also dictates that the scrubber be located for ease of access. The water in the scrubber can freeze and usually does if the system is not located in a heated blower penthouse on the roof or top floor of the laboratory building. The de-mister system must be efficient so that you are not exhausting hail pellets during the winter and acid rain during the summer.

At a recent meeting of engineers I heard several remarks to the effect that almost all hoods need scrubbers to protect the environs. God forbid. I do hope that some environmentalist in local, state, or federal government does not decide to save the world by requiring scrubbers to be mounted on all hood exhaust systems. If we would review these units you would find that 50 percent were totally useless; 49 percent were not needed and only 1 percent were being effective. The costs involved in such a project would pale our national debt.

INDICATING MANOMETERS

For years I have been preaching the gospel that demands an indicating device on all fume hoods so that the user knows that it is drawing the proper amount of air. I felt vindicated in my belief when OSHA [1] included this little rule in their latest revision of the regulations for chemical laboratories.

When you tour new facilities, and even some older ones, you see a wide spectrum of instruments that are intended to show a hood operator the status of the exhaust air from his or her hood. These can range from very sophisticated digital read out instruments with color coded lights, simple dial or inclined manometers or perhaps just a piece of tissue taped to the bottom of the open sash.

For reliability, simplicity, and economy the indicating manometer is pretty hard to beat. They have hardly any moving parts, no electrical or electronic circuits, and they are simple to mount on the hood face and to connect to the exhaust duct. The manometer shows the hood user that there is a certain pressure drop when an adequate volume of air is moving through the hood and into the exhaust duct.

The manometer should be mounted at eye level on the face of the hood where the user can readily see it at all times. It is connected to the exhaust duct using standard laboratory flexible plastic tubing and a simple plate and metal tube configuration. This is screwed on the exterior of the ducting after a small hole has penetrated the duct proper. The metal tubing can be soldered or welded onto the metal plate depending on the material being used. Refer to Figure 4.6.

When the fume hood exhaust system is first put into operation and the face velocity is set to the proper (corporate designated) value, the pressure drop should be marked on the manometer sight glass using a marking pen

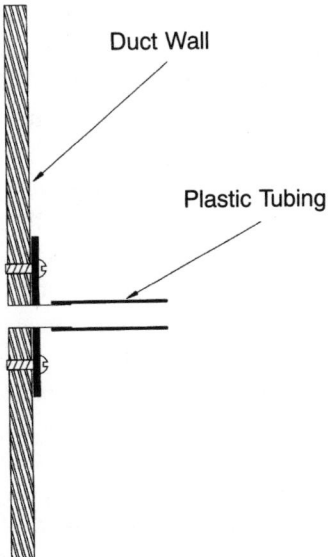

FIGURE 4.6 *Duct pressure tap for an indicating manometer.*

or a small brush with a quick drying paint. As long as the manometer needle is on or close to this line, the face velocity (exhaust volume) should be as it was originally established.

As time passes duct systems get dirty and may develop small (or large) holes, fan belts get frayed and slip, and the air volume will gradually decrease. The manometer will show the operator that this is happening by a drop in the indicated pressure and the building maintenance staff should be alerted. If the pressure decrease is very rapid, cease work in the hood, stop your experiments, and call the safety or industrial hygiene staff for immediate assistance.

The indicating manometer works only if the hood user pays attention to what it is indicating. Pay little or no attention to it, and it is useless.

ALARM SYSTEMS

Since I am pounding the drum for the manometer family I will state up front that you can very easily have an audible and visual alarm system using the manometer by adding electrical and/or electronic circuitry. Some manometers are manufactured with adjustable alarm set points as integral parts of the instrument. Other manometers can have adjustable subassemblies added to establish an alarm set-point system. To see what is available I suggest that you contact Dwyer Instruments, Inc., P.O. Box 373, Michigan City, Indiana 46360 for a catalog; they are the principal manufacturer of this equipment. You can go well beyond the simple dial manometer as such; there are digital read-out units, those with sequencing LED light patterns, and I am sure there are new products that I have never seen. Review them all and select the type that is best for you.

Other types of low-flow alarms, not based on static pressure drops, are also readily available. You will find hot wire sensors that are easily mounted in the hood side walls. There are other very sensitive instruments that sense the difference in pressure between the room and the hood chamber and translate this into face velocity. VAV systems, if you are fortunate enough to have one as part of your hood control instrumentation, are all designed to indicate and alarm for a low-flow situation.

As time passes I am sure that newer and better systems will be developed by the instrumentation fraternity. I would only ask that these companies determine what values should be indicated by their equipment. Currently, there is one company that tells the user that the air flow is in the energy efficient range, whatever that may mean. I want to know that my hood face velocity is safe, not energy-efficient. A little cooperation between the manufacturer and the user would be a great help to all.

A large number of alarm systems allow for the set point to be adjusted by the user. This is a very bad idea. The laboratory worker should know that his or her hood has an adequate face velocity. But the alarm set points should be determined and set by the safety staff. A researcher has enough to think about without getting involved with hood face velocity and pressure drop settings; both of which they probably do not fully understand. All alarms should be both visual and audible. The audible sound should have a "locked-in-position" silencing switch and only the supervisor should have the key; the visual component should remain lit until the low-flow situation is corrected.

All laboratory personnel from the janitor to the director of research must be aware that when a malfunction occurs it must be reported to the safety group immediately. Once a hood is not safe due to the lack of sufficient air exhaust, it must be closed to any continuing procedures until the air flow can be properly restored. If the failure occurs during the critical part of an experiment or process, some hard decisions must be made and people alerted. DO NOT TURN OFF THE AUDIBLE ALARM AND CONTINUE WORKING. You could very well endanger yourself and all the other occupants of the room. The low-flow alarm says the hood is unsafe to use. Heed the warning.

MATERIALS OF CONSTRUCTION

The principal item that allows for a large choice of materials is the hood liner. The materials range from stainless steel to cement board to various plastics. Each can have a special place for hood liners and new materials are being presented all the time. The hood manufacturers do not produce these lining materials; they buy them and cut them to the proper size for their products. This gives each manufacturer access to the same materials as a competitor. This means that the hood customer is not denied any type of material because of the choice of the company that might make the hoods.

To select the proper lining material for your usage, have the hood fabricator supply you with the specification documents for all the various materials; this will include both the chemical resistance schemes and the physical properties. It is wise to select a material that provides you with a white, or at least a very light (color-wise), interior hood finish. Be certain that the finish is resistant to your reagent usage and preferably is not too glossy. Hood chambers that are dull and dark are really depressing and become difficult to face each working day. The material should be reasonably smooth and easy to clean. If it is a painted finish, it should be capable of touch up or repainting in-place.

To minimize rusting problems it is wise to eliminate as much exposed metal from the interior of general purpose hoods as is possible. Plastics, if compatible, are good choices. Special coatings over steel are equally as good as long as the finish is extremely resistant to both chemical attack and physical abuse. I mentioned earlier in Chapter 3 that I had a preference for Kynar coatings. There are or will be others and it is wise to keep testing proposed finishes as they come on the market. Hood manufacturers can provide a lot of data for you but they have to be pushed and prodded to provide you with the latest technical data and the associated cost of the coating process. The usual ploy by fume hood salespeople is to talk you into the use of stainless steel in lieu of special coatings and/or plastics. If this happens to you, choose a new supplier or at least another salesperson because this type of presentation shows a low level of expertise. Remember what we said earlier: "Stainless ain't."

REFERENCE

1. 29 CFR 1910.1450, U.S. Government Printing Office, Washington DC, (1990).

5

Face Velocity

For impact, I am going to repeat part of the Introduction of this book with some of the historical events that have shaped today's understanding of fume hood face velocities.

Fume hood face velocity has been a very lively and controversial subject for health professionals and engineers since the 1930s and 1940s. In the early days of the Atomic Energy Commission (AEC) a hood face velocity of 50 fpm was considered adequate and safe. These were also the years of a widely and wildly expanding hood population.

Advanced hood design was just beginning in the 1940s and many questions were arising as to what was a safe face velocity. Unfortunately,the "more is better" fraternity became very vocal and convinced the majority of fume hood users that much higher face velocities were safer. At that time air velocity was the only parameter that could be measured objectively. There was some use of smoke wands and smoke bombs, but the results were more subjective than objective, and the face velocity was the primary focus of so called hood safety.

As the 1950s, 1960s and 1970s came and went, these "experts" were increasing the face velocity as rapidly as possible. From 50 fpm they had gone to 75, 100, and finally to 150 fpm and higher before some calmer heads prevailed. Some members of the fume hood community recognized that the increased velocities had not fostered safety and were driving the cost of building and operating laboratory buildings out of reach. In the late 1970s the American Society of Heating, Ventilating and Air Conditioning Engineers (ASHRAE) instituted a monumental research project [1]. Dr. G. W. Knutson and K. J. Caplan, undertook a project to establish a method of quantitatively testing fume hoods. We will study their report and resulting protocol in Chapter 9, but for now let us concentrate on a cursory overview:

1. Face velocities in the 60 to 100 fpm range provided acceptable and safe hood operating conditions.
2. Room air patterns can account for at least 50 percent of unsafe hood performance.

These data were reviewed and then published in 1982 by the American Conference of Governmental Industrial Hygienists (ACGIH) [2]. For the first time there existed a defined statement regarding hood safety and face velocities that was promoted by a society of health professionals and substantiated by excellent research.

It is only now (1993) that this information is being widely disseminated. Some manufacturers still fall into the trap of the 1960s and 1970s and rely on face velocity as a safety criteria rather than hood and room ventilation design. Some manufacturers' catalogs still reflect parameters as outlined in SAMA Standard LF 10-1980 [3], where so called class A, B, and C hoods were represented with face velocities up to 150 fpm. In view of more current data, these A, B, and C classifications and their implied face velocities should totally be ignored.

Set your hood face velocity based on the latest data and on factual rather than emotional presentations. I sat in a meeting with the design staff of a very large corporation and the staff industrial hygienist made a great statement: "We work with nasties so we will use 150 fpm as the design velocity." It is difficult to assess what that decision cost the company in increased costs and decreased safety. Whatever velocity you choose is up to you, and you must defend it to both the laboratory and the accounting staff. If you have chosen more than 100 fpm, go back and reevaluate your decision, or more likely the decision of an aging vice president or director of research.

To effect savings of the exhaust air, and thereby monies, many facilities are adopting policies that the fume hoods should be used with the front sash in a partially closed position and the volume (velocity) established with this criterion in place. This is not all bad but it sure can present a bag of worms if all the facets of the policy are not examined and recognized.

Regardless of the checks and balances used to require that the hood is actually used at the partially closed (or open) position, they will fail at least 25 percent of the time, and this figure may be as high as 50 percent depending on the integrity and reliability of user discipline. Laboratory personnel are not used to working with a partially closed sash and it will take both time and training to ensure compliance.

To be on the safe side, the measured face velocity of the hood, at a partial opening, should be close to what the velocity should be with the sash fully open. Remember the ACGIH set 60 fpm as an accepted minimum. As an example, if we have a hood with a 30-in. available sash opening, but want to use 18 in. as the restricted size opening then the ratio of 18 to 30 in. is 60 percent so the velocity at the 18-in. opening should be 60 fpm/60% or 100 fpm. If you are not comfortable with a minimum velocity of 60 fpm, you will have to insert a value that you prefer. Remember that overall sash opening height varies with manufacturer and with model, so you have to accommodate whatever the overall size actually is.

One last comment: Let us say that you choose 100 fpm as your minimum value; then when the sash is partially closed the measured velocity will be close to 170 fpm. Someone says it is too high because they have read that face velocities above 150 fpm can in themselves be detrimental to hood performance. The hazardous 150 fpm that they are referring to is the one present when the sash is full open. Except for sucking an occasional wipe up and out of the hood, these higher velocities become less and less of a disruptive factor, regarding operating performance, when the sash is moved to a smaller partially closed position. We discuss partially closed sash operation in Chapter 7.

REFERENCES

1. K. J. Caplan and G. W. Knutson, "Laboratory Fume Hoods, a Performance Test," RP 70, *ASHRAE Trans.*, 84, (I)(1978).
2. *Industrial Ventilation Manual*, 17th edition, American Conference of Governmental Hygienists, Cincinnati, 1982.
3. Laboratory Fume Hoods, SAMA Standard LF 10-1980, Scientific Apparatus Makers Association, Washington DC, (1980).

6

Systems Design

ROOM EXHAUST AIR VOLUMES

Using the hood face velocities and room heat loading (Chapter 5), the mechanical engineer will set the room air volumes. This will be based on the maximum velocities to be used for the hoods when the sash is fully opened. If you want to reduce the air volume so as to initiate savings in both construction and operating costs, you can do some of the things we have briefly mentioned and some items we have yet to consider.

We covered the selection of hood face velocity based on knowledge. Horizontal and horizontal/vertical sashes were diagrammed so as to save precious air. Now we shall consider further methods in this chapter on systems design.

Simple suggestion: Turn off the hoods when not in use. This is much more complex than just adding an on/off switch. When doing this you must bear in mind that it is the supply air you must reduce as this is the air that is costing you money. Laboratory space must remain at a negative pressure in relation to the corridor and service space to prevent outward air migration. This requires a damper in the room supply air system so this make-up air can be controlled in relation to the hood exhaust. To further complicate the system, the heat loading may very well require more air volume than can be obtained from the hood itself so you have a secondary exhaust outlet that in turn must be modulated. This modulation must be rapid and positive so that the laboratory space does not go through cycles of positive/negative status. In a central exhaust fan system you have either pressure or volume or combination sensors that regulate in-line dampers. In a single blower per hood configuration, you acquire an added problem of having to close off the hood exhaust blower ducting so that there is no recirculation back into the lab through the idle blower and its associated ducting.

Another item to consider is that you should not turn a hood completely off. In most installations the cabinets under the hoods are used for acid and/or solvent storage and require a slight through-put of air. On single fan per hood systems you should have blowers with two speeds and this can be accomplished with dual windings or a frequency modulation control. In a central exhaust system you accomplish the turn down with in-line dampers. Remember one thing: in-line dampers are a source of noise, and there is little you can do about it.

VARIABLE AIR VOLUME SYSTEMS

The latest type of instrumentation on the market is variable air volume (VAV) control. This approach regulates the exhaust from the hood so that you have a fairly constant face velocity, with the implementation of a varying volume. These systems are quite sophisticated and track the make-up air to the total exhaust by means of multiple dampers. The good systems react instantly to changes and by and large do a good job. There are several manufacturers of VAV equipment and they utilize one of three recognizable techniques:

1. Velocity sensing in the hood side wall.
2. Sash position determination.
3. Pressure sensing between hood and room.

At this time the sash position sensing offers the best and most accurate control. However, this industry is in its infancy and new and better products reach the market on a continuing basis. If you are going to evaluate this approach, and you should, make it a precondition for evaluation that you visit some on-stream installations; preferably ones having at least a year of operating experience. In selecting a particular brand, thoroughly investigate the field help that is guaranteed by the manufacturer. It has been my experience that if the systems are installed and calibrated properly they offer a minimum of trouble. They scare many an in-house maintenance crew but they really shouldn't. Remember how complex a TV set is and yet we all have them. We also know who to call when they quit. In the more reliable VAV systems the electronic components are plug-in boards and have a built-in diagnostic system for quick trouble location.

These VAV systems can have an analog, a digital, or an analog/digital control presentation. Those with digital output can feed into any one of a variety of building control systems. If properly programmed into a computer archiving program, you can recall the statics or dynamics of a single hood or the entire hood system at anytime in the memory span of the stored data. This feature would be extremely useful if you have to provide a history for some unexplained ventilation excursion or accident in the building.

I have been to seminars and have listened to manufacturers, consultants, and users discuss VAV systems and I have been favorably impressed. I was pleased with the inherent system safety. To have a preset and somewhat accurate hood face velocity is the basic protection. Two other factors contribute to added safety: (1) An alarm when the ventilation system is going down and (2) the ability to go to the full exhaust potential when the sash is closed for emergencies such as a fire or a bad toxic spill in the hood.

VAV systems do not come free. The current estimate is around $5000 per hood. The claims made by some of the consultants and suppliers would indicate a payback in months. This leads me to a lot of questions and skepticism. I think that there are a lot of concurrent factors that must be present within the system to provide a good payback and this return on investment is not solely derived from the VAV controls. My feeling is that you must also have (1) a reasonably large hood population. (2) good user discipline so that the hood sashes are in their lowest operable height when in use and that the sashes are closed when the hood is not in use, and (3) some type of off-hour set-back design so that the ventilation system can go into a reduced volume program when the labs are not occupied. When you look at these three points of reference they should be part and parcel of every hood training program whether you are using VAV controls or not. It takes a lot of common sense to make anything or everything work.

The ASHRAE meeting that was held in Vancouver, British Columbia, in 1989 furnished some very interesting papers on VAV systems [1, 2]. But as I said before, the world of instrumentation is in a constant state of flux and really requires that you always review the latest product available and judge for yourself. Also, take conducted tours of working labs with the head of maintenance not the director of research.

VAV users should be reminded that as the sash is closed on a hood, the amount of dilution air diminishes. You may not be able to get below a lower explosive limit (LEL) with this reduced flow.

AUXILIARY AIR FUME HOODS

An auxiliary air or supplemental air fume hood is one that is designed so that a source of air, from outside the room proper, is introduced near the top and front of the hood sash so that the amount of room air exhausted by the hood proper is reduced. As with all types of equipment, there is "good, better, and best." Likewise, there is the right way to incorporate them into a system and the wrong way. I hate to say it, but with the thousands of auxiliary air hoods that are currently in use, most of them are improperly incorporated into the ventilation system. The two primary objections are from the users: "I freeze to death in the winter and get a stiff neck." "It gets so hot and humid in the laboratory in the summer that it becomes unbearable." This is not the fault of the design engineers because the manufacturers insisted that the supplemental air could be of very inferior quality as compared to the room air. They recommended heating capacity in the ducts to raise the auxiliary air to 55° F in the winter and bring it in untreated from the outside in the summer. I don't care what anyone says, the supplemental air must be very close to room air quality for the comfort of the laboratory staff.

Getting back to the "good, better, best" classifications, some designs do a vastly superior job of presenting the supplemental air to the fume hood than others. First, the auxiliary air must be of a reasonably uniform velocity (plus or minus 20 percent) across the discharge area of the auxiliary air chamber. Second, the size of the discharge area must be such that the discharge velocity does not exceed the face velocity of the fume hood by more than a multiplying factor of 1.20. Lastly, the noise that the auxiliary air chamber generates must not be objectionable to the users. If you are considering incorporating this design into your facilities, go visit other laboratories and judge the various manufactured products with these three checkpoints in mind.

Auxiliary air design can do a superb job of providing a safe hood in which to work. This is particularly true for walk-in hoods. In the chapter covering quantitative testing (CHAPTER 9) we will see a test profile of a hood with and without the auxiliary air and the difference is remarkable.

Maintenance departments claim that to balance the air volumes for this style hood and then to keep them in balance are tough jobs. This could be true because if you have two quite large central air handling systems with different aging factors they can become unbalanced. I have not seen any VAV controls applied to these systems, but if some bright design engineer can merge the two you could find them both energy efficient and safe to use.

Many times the argument is put forth by design people that because supplemental air requires both a supply and an exhaust system that they are not capable of reducing initial capital or ongoing operating costs for a laboratory. This is not necessarily true. In Figure 6.1 we have a computer generated model of a laboratory building using standard ASHRAE procedures and formulas to set system costs. Let us set the parameters for the model:

1. Medium sized building with 20 laboratories.
2. Each laboratory has two 6-ft fume hoods.
3. The hoods operate 10 hours/day, 220 days/year.
4. The hood diversity (see below) is 75 percent.
5. Ninety percent first-pass capture efficiency (see below) for the auxiliary air.
6. Auxiliary air is room temperature in the winter, 15° F above room temperature in the summer, with the moisture content reduced to 80 grains per cubic meter.
7. The building is located in Chicago for temperature and humidity calculations.

In this model we are showing two types of sash, horizontal and vertical; four relative quantities of auxiliary air, 0, 50, 70, and 95 percent; and each data set shows operation at three different hood face velocities: 60, 80, and 100 fpm. First-pass capture efficiency (refer to parameter 5 above) is defined as the amount of auxiliary air that is captured by the fume hood on its first pass by the hood face and the balance, or sluff-off, is considered as room loading. See Figure 6.1. Note the amount of both construction and

Laboratory Setup		Hood cfm	Auxiliary Air cfm	Total Tons/Lab	HVAC Costs/Lab	Auxiliary Air Costs/Lab	Total Initial Costs/Lab	Yearly Operating Costs/Lab	Initial + 5 Yr Operating Costs/Lab
60 ft/min Face Velocity									
Vertical sash-full open	0% Auxiliary air	1620	0	9.31	$12,442	$0	$12,442	$571	$15,295
Horizontal sash-50% open	0% Auxiliary air	810	0	5.32	$7,768	$0	$7,768	$316	$9,350
Horizontal sash-50% open	50% Auxiliary air	810	405	3.74	$4,960	$2,182	$7,142	$272	$8,504
Vertical sash-full open	50% Auxiliary air	1620	810	6.68	$8,104	$3,864	$11,968	$498	$14,457
Vertical sash-full open	70% Auxiliary air	1620	1134	5.84	$5,952	$5,209	$11,161	$475	$13,535
Vertical sash-full open	95% Auxiliary air	1620	1539	4.79	$3,263	$6,891	$10,154	$446	$12,383
80 ft/min Face Velocity									
Vertical sash-full open	0% Auxiliary air	2160	0	11.97	$15,557	$0	$15,557	$740	$19,258
Horizontal sash-50% open	0% Auxiliary air	1080	0	6.65	$9,326	$0	$9,326	$401	$11,331
Horizontal sash-50% open	50% Auxiliary air	1080	540	4.72	$6,008	$2,742	$8,751	$347	$10,488
Vertical sash-full open	50% Auxiliary air	2160	1080	8.64	$10,200	$4,985	$15,185	$648	$18,425
Vertical sash-full open	70% Auxiliary air	2160	1512	7.52	$7,331	$6,779	$14,110	$617	$17,197
Vertical sash-full open	95% Auxiliary air	2160	2052	6.13	$3,745	$9,021	$12,766	$579	$15,661
100 ft/min Face Velocity									
Vertical sash-full open	0% Auxiliary air	2700	0	14.63	$18,673	$0	$18,673	$910	$23,222
Horizontal sash-50% open	0% Auxiliary air	1350	0	7.98	$10,884	$0	$10,884	$486	$13,313
Horizontal sash-50% open	50% Auxiliary air	1350	675	5.70	$7,056	$3,303	$10,359	$423	$12,472
Vertical sash-full open	50% Auxiliary air	2700	1350	10.60	$12,296	$6,106	$18,402	$798	$22,394
Vertical sash-full open	70% Auxiliary air	2700	1890	9.20	$8,710	$8,349	$17,058	$760	$20,858
Vertical sash-full open	95% Auxiliary air	2700	2565	7.46	$4,227	$11,152	$15,378	$712	$18,938

FIGURE 6.1 Computer-generated model for systems costs.

operating monies that can be saved by using horizontal sliding sash in lieu of a vertical rising sash system.

HOOD DIVERSITY

Hood diversity is not mentioned as much as it should be and sometimes this factor is used improperly. Hood diversity is the percentage of hoods in a laboratory building that are being used at the same time. A 75 percent diversity means that 75 percent of the hoods are turned on simultaneously, whether they are actually being used or not is an academic question.

HVAC design engineers frequently use a 75 percent diversity factor in sizing their total building air volumes. Whether this is accurate or not depends on the discipline of the hood users. I know of one very large midwestern research laboratory, which had an excellent computer record of hood operation and equally as good discipline, that showed a diversity of 41 percent. Design engineers face somewhat of a dilemma in using the diversity factor because of anticipated user discipline. Engineers can underanticipate and this can cost a bundle and a half to upgrade a couple years down the road. Likewise, their estimate can be too high and they spend an excessive amount of money on the initial construction. If the design and user staffs would approach the use factor with both feet on the ground, and not floating on some cloud, the initial construction costs and the ongoing operating costs could be reduced without jeopardizing hood performance or worker safety [3].

BLOWER SYSTEMS

When faced with problems, design engineers become most resourceful and inventive. New damper and mixing boxes came on the market, followed by a revolution in sensing and control instrumentation. New and better replaced the old and tired. Systems became totally integrated; exhaust stack height and building orientation are calculated by new environmental models. Exhaust air reentry is recognized in a different light and it now is a systems evaluation and not a blower by blower investigation. Today, except for isolated installations, you rarely see the one hood/one blower building exhaust system. This improved technology has made buildings more acceptable and safer and it is quite feasible to use a central type exhaust system even for small installations. It does require that the HVAC group be on their toes and know what advances have come along and not just copy some type of profile they have been using for years just because they are comfortable. If this group does not challenge itself, then some other group must. While I do not know all the questions to ask, I do know that if you start with the basic ones then the others will evolve. The message here is to question yourself or the design group long before the concrete is poured.

HOOD NOISE

"My hoods sure do make a lot of noise." This is not an unusual statement from the laboratory staff. Most of the blame is usually heaped upon the head of the hood manufacturer. Let it be known up front that hoods in themselves do not generate noise. They are like the headphones connected to a portable tape player; the player provides the sounds and the earphones transmit it to the listener. The hood is therefore only sending out the noises generated by the exhaust system. These system noises come from a variety of causes: air going in excess of 2000 fpm thru the ducting; turbulence caused by the improper treatment of ducting turns or transitions to smaller or larger ducting; unbalanced blower impeller wheels or frayed and cracked blower/motor belts; assorted junk left in the ducting during construction. In-line dampers cause noise when they are partially closed. There are probably more reasons than one can imagine and ventilation noises are very noticeable because most laboratories are reasonably quiet. I have found while investigating hood system noise that the audible component is more likely caused by a number of small factors than by one large problem. Investigating and curing exhaust system noise are a slow and patient process, and the final cures are not always expensive.

Fume hoods can amplify the noises that are generated by the exhaust system. There was a most interesting and informative article in *American Laboratory* that addressed this subject [4]. Dr. Robert Haugen, of St. Charles Manufacturing Co., tested for noise generation from an operating duct stub. He then measured for noise amplification and attenuation when a bench hood was connected to the duct stub.

The ducting and blower system (in this particular installation) generated a noise pattern that peaked between 500 and 2000 Hz. When the hood was connected, to the stub, the noise pattern, with the sash either open or closed, produced additional noise levels in the range of from 40 to 1500 Hz. The overall dBA rating of the duct stub alone was 64.5 dB and when the hood was connected the noise level rose to 67 dB. He demonstrated that the noise level increase was a direct function of hood size (width). The stub alone has a noise radiation angle of 180°, with the addition of the 6-ft hood the angle was reduced to approximately 100°. The maximum amplification over the baseline duct noise occurred at a wavelength close to the interior width of the hood cavity.

If you have hood noise, don't totally blame the hood manufacturer. Go to the HVAC design group and make them find the causes and furnish the corrections. This group is fully capable of defining and correcting the problem.

REFERENCES

1. T. M. Rabish, et al., "Comparisons of Variable-Volume Fume Hood Controllers," *ASHRAE Trans.*, 95, pt. 2 (1989).
2. D. R. Lacey, "Observed Performance of VAV Hood Controls," *ASHRAE Trans.*, 95, pt. 2 (1989).
3. R. C. Moyer, "Fume Hood Diversity for Reduced Energy Consumption," *ASHRAE Trans.*, 89, 2A and 2B (1983).
4. R. Haugen, "Resolving the Controversy Over Fume Hoods and Noise," *Am. Lab.*, 28 (November 1991).

7

Discipline

Laboratory people, by the very nature of their work, should be highly disciplined when it comes to all their laboratory operations. If this means keeping equipment or product clean, they do an admirable job; for the proper dilutions of chemicals they do very well; however, when it comes to the proper use of their fume hoods the word "discipline" normally goes out the window.

You can see hoods piled high with all sorts of apparatus, some being used and some useless. Some hoods are used as a storage area: a wall-to-wall collection of bottles, cans, and boxes. I've seen the low-flow alarm system disconnected because the constant buzzing of the warning alarm annoyed the user. This only scratches the surface for breaches of discipline; but we really need to find out WHY we have these violations of safety while other laboratory functions are carried out almost flawlessly.

I do not think that the laboratory staff is ignoring hood safety just to be ornery. I truly believe that they just do not know any better. They have never received even the most rudimentary instruction on hood usage. Perhaps freshman chemistry should devote a minimum of 3 hours of class exposure to fume hoods. If you are a betting person I offer to wager that there are, at the most, three institutions of higher learning giving any basic hood safety instruction; maybe none. With at least 3000 junior colleges,

colleges, and universities in the United States this is indeed a sad commentary.

A more visible number of corporate and governmental laboratories do attempt to have regular safety meetings in which fume hoods are included. The current OSHA regulations for laboratories [1] address fume hood safety and demand a comprehensive training program for all laboratory personnel. But even with all this attention, the proper techniques do not receive adequate recognition because most of the people that make these presentations on fume hoods are poorly informed. The OSHA personnel that should monitor and enrich these sessions would be hard pressed to contribute very much substance to the program. I would again place a wager that, at the very most, only one person on the OSHA staff really understands how hoods operate; maybe none.

To the trainers and to the trainees I offer the following: know the basics and the rest will come as you question and learn.

For our exercise in learning let us set up a series of questions that will cover the basis for acquiring better discipline, then when you search for the answers and the reasoning behind the answers things will start to fall into place.

1. Is the hood face velocity adequate?
2. Are the back baffle slots properly adjusted?
3. Is the work 6 in. back into the hood chamber?
4. Is my hood housekeeping good?
5. Is the bottom front air foil in place?
6. Does my sash slide easily?
7. Do I keep the vertical sash closed to a point below my shoulders when I use the hood?
8. Do I keep my horizontal sash partially closed so as to obtain the smallest possible, but reasonable, opening.
9. If my hood has an indicating manometer or a low-flow alarm, do I know how they work?
10. Can I rely on the noise I hear when standing at the hood face to tell me that the hood is working properly?
11. What do I do if I have a fire in the hood?
12. Have my health or safety departments performed a qualitative or quantitative evaluation of my hood, and if so, WHEN? What were the results?

Here are my answers to these questions.

1. As I have said many times throughout this book, and as an echo to the latest published data, the average hood face velocity should be no lower than 60 fpm nor any higher than 100 fpm when the sash is in a full open position. (See Chapter 5.)

2. The back baffle slots should be set so that the top slot is no larger than $\frac{1}{2}$ in. open and the bottom slot is between 2 and $2\frac{1}{2}$ in. open. These settings give the necessary "floor sweep" with minimum "vortex" generation. (See Chapter 3.)

3. DuPont and many other research oriented companies have found that procedures carried out at least 6 in. back into the hood have a much higher level of safety than those performed at the very front. This is because the room air patterns have much less effect on work carried out in the area of the hood that is more than 6 in. back from the hood face. (See Chapter 2.)

4. This commonsense question has a very commonsense and obvious answer. The more "stuff" that is crammed into the hood, the poorer the internal exhaust air patterns. The air is impeded and this will limit the flow of air away from your work and out of the hood exhaust. This is particularly true if you block the bottom back baffle slot. You solve a great part of this problem by only having what you need for current work located inside the hood, and whatever this may be, it is supported off the hood work surface about $1\frac{1}{2}$ in. using rubber stoppers, baby food jars or pieces of $1\frac{1}{2}$ in. PVC pipe. (See Chapter 3.)

5. I have seen the bottom front air foil removed from hoods because the user said it was in the way. That is like removing the brake pedal in your car because it is where you want to rest your foot. This air foil is a major part of hood design and should never be removed from a working hood. (See Chapter 3.)

6. The safety-glass hood sash may be your very best friend if you need a place to hide in the case of a quick fire or explosion. I have tried to move the vertical sash on some hoods and found that no amount of force would do this. The sash cables had slipped off the pulley system and they had been this way for months. Or there were obstructions in the horizontal track that blocked the sash travel. (See Chapter 3.)

7. As the sash is lowered there is less area in the front plane of the hood that can be influenced by cross drafts or traffic problems. (See Chapter 4.)

8. Again, with the closed sash to its smallest, but usable, opening, you

will have less influence on hood performance from room air currents and traffic patterns.

9. An indicating manometer installed on a hood is meant to show that the system is operating properly. Whether it does or not depends on how it was set when the exhaust system was new. There is a direct correlation between the exhaust volume and the pressure drop as shown by the manometer. In general, a higher reading indicates that a greater volume of air is flowing through the ducting. The low-flow alarm is basically a manometer that has an audible and visual signal when the settings are violated by poor performance. The hood user must know how these two items were calibrated, set, and operating. It would be like checking your fuel gage before starting out for Boston. (See Chapter 4.)

10. Because you hear noise when you are standing in front of your hood is no guarantee that it is working, let alone properly. The noises could be coming from a variety of sources in the exhaust system, but not necessarily from your hoods' system. Current OSHA regulations require that each hood have some sort of device that shows the user that the hood is working properly.(See Chapter 4.)

11. A fume hood is an ideal place to have a laboratory fire, if indeed you must have one. The hood is a highly ventilated area and is made of fire-resistant materials. If you have a flash type solvent fire, you might just lower the sash and wait a few seconds for it to burn out if the volume of the solvent is not great. If the burning reservoir is fairly large you may choose to use a CO_2 or dry powder extinguisher before calling for outside help. Never use water on a solvent fire because it will cause the fire to flow out of the hood and into the laboratory and then you will really have a problem. If the hood has a VAV control, override the system with the emergency button on the controller, so that you can have full exhaust volume with the sash lowered. If you have an equipment fire that you do not feel you can extinguish, pull the sash down, activate the building fire alarm, exit the room, close the door, and wait for professional help. (See Chapter 4.)

12. We should ask this question for two reasons. Either the testing has been done and the data are readily available, or the responsible staff group gives you a stare indicating that they don't know what you are talking about. Current OSHA regulations [1] require that a laboratory be fully aware of the operating status of all its fume hoods. If your hood has not been qualitatively and quantitatively tested, then both you and the cor-

porate structure could be in for some trouble. Frankly, I'd be more concerned about number one (myself) and tell the facility staff that they must do viable testing and with favorable results. (See Chapters 8 and 9.)

What we have reviewed is really attitude and the approach toward hood safety. Only you can be held responsible for your discipline and only then if you do or do not know better. To the individual I say "learn"; to the health and safety groups I say "teach." To both of you I would point out that good hood discipline could very well mean the difference between being healthy or sick or perhaps even disabled.

REFERENCE

1. 29 CFR 1910.1450, U.S. Government Printing Office, Washington DC, (1990).

8

Qualitative Testing

The chapter that follows this is dedicated to the quantitative protocol for testing hoods. You may feel that you could be wasting your time on the qualitative tests of this chapter when you could jump to the real world and start on the quantitative tests. By so doing you could be wasting a lot of time and money. The qualitative tests point to the areas of poor hood performance and the quantitative testing is the focus to identify the specific causes. We will call this qualitative overview "screening."

A large percentage of hood problems can be detected, and solved, with the normal instruments and devices that every safety or occupational health group should have on hand:

1. A calibrated velometer.
2. Smoke sticks or a bottle of titanium tetrachloride.
3. A supply of 30-second smoke bombs [1].

Just as a comparison, quantitative testing requires much more sophisticated instrumentation with associated higher costs (approximately $15,000) and the requirement to spend considerably more time per hood than does qualitative testing.

To acquaint ourselves with the basics of "screening" we discuss the equipment listed above.

There are some very good velometers now being marketed. They can vary from the vane velometer to the propeller type and ultimately the hot wire thermoanemometer. The vane type instruments have a $200 to $600 price range and the propeller types fall into this price category also. The hot wire instruments range from $350 to $1350. The vane instruments have three major problems: they fluctuate widely while taking readings; you must be on the right side of the instrument while taking readings so that you can read the scale; they are large enough so that they do not fit into a lot of hood areas in which you would like to take readings. The propeller type gives an averaged reading without indicating to the user the extent of fluctuation, this is a real handicap because you really should be interested in seeing the extent of the turbulences in or near the hood. The propeller units are not as inflexible as the vane type instruments but they do run a close second. Obviously, I prefer the hot wire thermoanemometer. The sensing head is small, it is only semidirectional, and the readings can be electronically dampened so you can see the swing in the readings but in a well regulated frame of reference. Within the thermoanemometer family you have both analog and the digital read out instruments. To me the digital instruments, which are the most expensive, are the least desirable. Within their electronic circuitry they average the readings by imposing a data collection time frame and you actually see data that has occurred over a set period of time; this can run from 2 or 3 seconds up to 10 seconds. This makes it most difficult to see the extent of any turbulence and rapidly varying air currents, and both are a necessary ingredient of the screening process. The analog instrument has a dampened meter movement in the range of 0.5 second that regulates the wild swing of the vane units and lets you see the turbulence that the digital instrument precludes. I can do a much more comprehensive hood profile with an analog unit in one-third of the time as with a digital instrument. I still feel that the human brain is a pretty fair computer and people doing velocity profiles can sort out the real high/low swings themselves and get a better feeling for the air movements. Whichever instrument you ultimately choose, be sure to read the instructions and learn how to use it properly and then be religious in having it calibrated on a yearly basis.

In the next few years there could be a lot of discussion as to what velometers actually measure. With new, and nondirectional, sensors entering the marketplace some diverse opinions might arise on the accuracy of the measuring devices, and how well they indicate the true face velocity of hoods.

Smoke sticks or a bottle of titanium tetrachloride gives you a localized source of smoke that can easily be moved about the area under investigation. Rather than buying a bottle of titanium tetrachloride and using cotton tipped swabs I much prefer the smoke sticks as marketed by the Mine Safety Appliance Co., 121 Gama Drive, Pittsburgh, Pennsylvania. They are small glass ampules of $TiCl_4$ wrapped on one end with cotton batting. A small pair of pliers will easily crush the glass under the cotton and you have a smoke source that will last 4 to 5 minutes. These sticks are safe to carry, will not spill and do not drip the $TiCl_4$ on the floor, the hood work surface, or your shoes, which by the way creates an instant hole in the leather; all of which can happen with the bottle/swab approach.

Thirty-second smoke bombs are an invaluable testing device but they can be very cantankerous. Sometimes they don't light and almost every time that they do ignite, there is a small explosive jet when the smoke starts to flow. You can almost eliminate the nonlighting problem by tamping both ends of the smoke bomb a couple or three times on the hood work surface before you try to light it. To minimize the small explosive jet, find the side discharge hole in the smoke bomb and puncture the single layer of paper that covers it. The hole is easy to see because of the indent under the covering wrap. Before lighting, place the smoke bomb where you want it, point the hole toward the back of the hood and only then light the fuse. Before it ignites hold a 12 in., round or square, piece of metal or cement board about 6 in. above the now smouldering bomb. A handle on the shield is a great help. The bomb will ignite with a rapid effusion of smoke and the discharge of some gummy tar material that the shield will collect instead of messing up the fume hood. You might even consider lowering the sash part way during the lighting ceremony as an added precaution and protection for yourself. I've used hundreds of these bombs, have never been injured, but I had messed up a lot of hoods until I started using an aluminum shield.

DETERMINING FACE VELOCITIES

As the starting point for screening we must measure the hood face velocity profile. First, we establish a grid pattern across the hood face where we will be taking our readings. The Scientific Apparatus Manufacturers Association (SAMA) published in their Standard LF 10-1980 [2] the accepted grids for the various sizes and types of fume hoods. Shown in Figure 8.1 is a typical grid for a 5-ft bench hood. To utilize these grids you follow a set procedure (see Figure 8.1):

1. Measure the opening size, that is, dimensions A and B.
2. Establish the grid spacing with a maximum overall grid size of 12 in. × 12 in. in.
3. Take the velocity reading at the center of each grid.
4. Add the individual readings and divide by the number of readings taken; this provides the average velocity in feet per minute (fpm).
5. Multiply the average face velocity in fpm by the total area in square feet (A × B) for the volume rate in cubic feet per minute (cfm).

Your velocity will be more accurate the more readings that you take. However, you must consider that these readings are no more accurate than

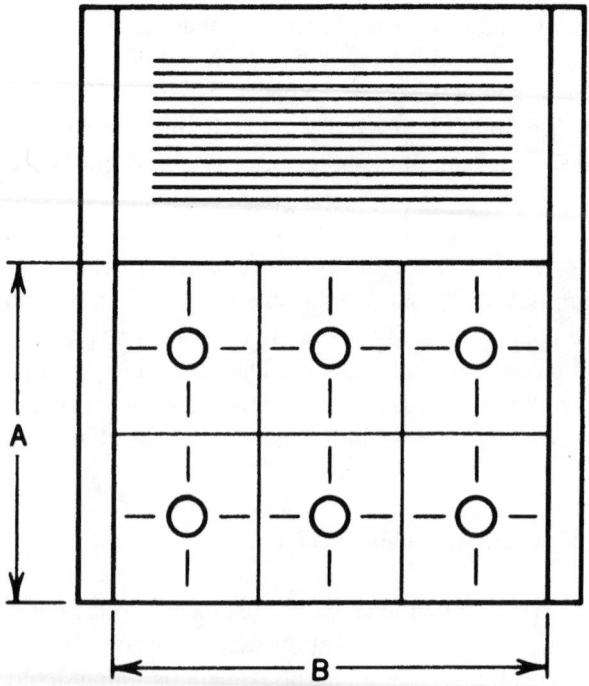

FIGURE 8.1   *Velocity profile measurement grid.*

plus or minus 10 percent. This degree of accuracy reflects the instrument, the user, and the fluctuation of the room conditions. Round out your readings to the closest whole number, as decimal point figures are useless. You should be equally as concerned about the distribution of the velocities as with the final averaged reading. Are the individual readings pretty stable or do they fluctuate across the scale from 0 to 100 and maybe even go into negative territory? If so, you could be experiencing some severe cross-drafts at the hood face. Are the readings much higher across the top portion of the opening than across the bottom? It is possible that your back baffle slots are improperly adjusted.

If all your grid readings are reasonably stable and are each within plus or minus 15 percent of the average velocity, I would not be too concerned about this pattern of velocities.

Now I am going to back up a bit and offer some practical advice that might help in achieving more accurate velocity profiles. The protocol for taking a face velocity profile is not defined in exact terms; suggestions yes, definitions no. Some people prefer to stand in the hood face opening when they are taking their readings. Others, like me, prefer to stand off to one side and not be in the hood air flow pattern. A vane velometer dictates that you must stand in the air pattern so that you can read the instrument. The thermoanemometer gives you the choice to go either way, as does the propeller type instrument. I stand out of the frontal area because I have convinced myself that I can get a less turbulent, and more accurate, reading. Standing off to one side or the other will produce slightly lower readings since the open sash area is greater than with a body present in the opening. Choose the method that is most comfortable and reliable for you and then always use it so that your results are consistent for your purposes.

After you are satisfied that you have a good velocity profile it is time to do a face traverse with a hand-held smoke device. The air should flow into the hood with a reasonably smooth pattern and no smoke should be flowing back into the laboratory proper. If it does flow back, you probably have at least one of three potential problems: poorly adjusted back baffle slots; poor housekeeping with a lot of equipment on the hood work surface without any spacers under them; or a bad room cross draft.

You are finished with the velocity profile and the smoke sticks, and all seems to be going quite well. We are now ready to do the most revealing of the qualitative tests. Following the directions that we outlined earlier in this chapter, we ignite our 30-second smoke bomb. The first bomb should be placed in the center of the hood and 6 in. back into the hood superstructure. Slowly raise the sash to full open and see if there is any visible smoke escaping into the room. A small continuous pattern, or even an occasional puff, would most likely indicate that you have an offending room air pattern or a laboratory traffic problem—but only after you had satisfied yourself that the housekeeping and slot adjustments were as they should be. It will take some patience and experimentation to pinpoint the exact cause. In Chapter 9 regarding quantitative testing, we will go through a complete exercise investigating cross-drafts caused by ceiling air diffusers. You might want to try it here with smoke patterns but only if the flow of smoke is very noticeable. I would really suggest waiting for the quantitative investigation.

DRY ICE AND WATER TEST

One last qualitative demonstration of fume hood performance is by the use of dry ice and water. For many years fume hood manufacturers have used a pie pan filled with warm water and some chips of dry ice to form a dense mixture of CO_2 and water vapor. This fairly dense combination provides an excellent visual demonstration of the necessity of floor sweep of a hood and the effectiveness of the bottom front air foil. Dr. John Adams, E.I. duPont in Newark, Delaware [3] refined the procedure with specific amounts of dry ice in water and postulated defined test results even comparing them to the ASHRAE 110-1985 protocol. While I do not agree with Dr. Adams' conclusions I do think that it is a very effective method of checking the floor sweep operation of a hood. My disagreement lies in the fact that this dry ice procedure only investigates floor sweep and ignores the vortex area of the hood chamber.

When you have completed this screening of your hoods, and have found no obvious problems, then the chances of you having a serious hood problem are relatively small. While we have screened the hood we are in no position to quantify the performance since we have only performed macro analytical procedures. The smoke concentrations that we can see are in the 200- to 300-ppm range and are subject to an individual's own judgment call. While the screening process does an admirable job of deciding on GOOD and BETTER performance, it cannot judge with any degree of accuracy BEST performance. This is why we will leave our simple procedures and instruments and proceed on to more sophisticated instruments and techniques.

Right now I will make another wager: at least 70 percent of the hoods that have passed the screening process will also pass the quantitative testing of Chapter 9. If they don't, you were either not consistent or not accurate in the screening process or you have a building air reentry problem that you were not able to detect with the smoke tests.

REFERENCES

1. Two sources of smoke-making devices are: E. Vernon Hill Co., Corte Madera, CA, and Superior Signal Co., Spotswood, NJ.
2. Laboratory Fume Hoods, SAMA Standard LF 10-1980, Scientific Apparatus Makers Association, Washington DC.
3. John B. Adams, "Synthesis Laboratory Fume Hoods: Easy Reliable Performance Evaluation; Importance of Sash Design," Stine–Haskell Research Center, E.I. duPont, Newark, DE 19714 (1988).

9

Quantitative Testing

At this time, quantitative fume hood testing and ASHRAE 110-1985 [1] (soon to be revised and republished as ASHRAE 110-1993) are synonymous. The procedures outlined in the ASHRAE Standard are now being recognized worldwide as the basis for determining safe fume hood performance. While I will not have a copy of ASHRAE 110-1993 prior to the publication of this book, I am in close touch with members of the Standards Committee and will include in this chapter a review of any significant changes made to ASHRAE 110-1985.

Since the publication of the results of the original research project (RP-70) in 1978 [2], through the proposed standard (ASHRAE 110-P) in 1982, and then the actual Standard in 1985, there have been suggestions in various journal articles for modifications to the original work.

In 1979 researchers at E. I. duPont followed a different reporting system [3]. Monsanto Chemical has chosen a different diffusion rate for the test gas. An EPA Standard [4] presents a vastly modified gas diffuser and an entirely different sampling procedure. Most recently, the British and European groups are proposing some modifications in the type and the numbers of the ejection diffusers used during the test protocol.

I have not tried to evaluate any of these changes as being either good or bad. I feel strongly that "if it ain't broke, don't fix it." I think that it is necessary to have one procedure and one reporting system so we can all compare fume hood performance regardless of the research group or the country of origin. Changes can be made to ASHRAE 110-1985 through periodic review by qualified people and mutual consensus. To make a change for change's sake is self-defeating. The Standard will be much stronger if everyone will work within its confines, bypass locally initiated variations, and keep in mind that the real purpose of the test is to develop safer fume hoods. If some group can support, by good fundamental laboratory work, a change that is indeed an advancement, fine; present the change to the ASHRAE review committee. Then a wide spectrum of qualified people can review the proposal, and it can be incorporated in a revised Standard if it is a recognized improvement. So much for my soap-box oratory.

The protocol as outlined by Caplan and Knutson (RP-70) in 1978 was a method to conduct a test. In 1982 the American Conference of Governmental Industrial Hygienists established a pass/fail rating system [5] when applying the proposed ASHRAE protocol. With this in mind let us now proceed with a review of the protocol that the Standard established.

The basic test procedure requires the placement of a three-dimensional clothing store manikin of average size (67 in. tall with a 16-in. shoulder width) in a location at the hood face similar to a person using the hood. The manikin is to be dressed in a lab smock, coveralls, or other reasonably close-fitting clothing. It is to be positioned so that the nose (breathing zone) of the manikin is at the front plane of the hood and not at the plane of the hood sash.

To the breathing zone area of the manikin you attach a collection probe that samples the air that would be inhaled by the hood user. This probe is connected to a continuous reading instrument that is specifically calibrated for the test gas being used. The range of detection of the instrument shall be at least from 0.01 to 100 ppm, with an accuracy of plus or minus 10 percent of the reading for concentrations above 0.1 ppm and plus or minus 25 percent for concentrations between 0.01 ppm and 0.1.

The tracer gas is usually sulfur hexafluoride; however, this may be changed depending on the mission of the laboratory in which the hood is being tested. For example, if the laboratory has been doing trace fluorocarbon analysis, it may not be able to operate while fluorocarbons are being released for hood testing. In such a case you may wish to change to a nontoxic, nonodorous, noncorrosive gas that can be analyzed by the detection instrument at the appropriate concentrations.

The trace gas should be in a pressurized cylinder that can maintain 30 psig at the selected flow rate for a period of 1 hour. Suggested flow rates are 1, 4, and 8 liters per minute (lpm). The tracer gas is piped to the ejector (see next paragraph) and must include a pressure gage and a positive shut-off valve.

The gas ejector is described with full construction drawings in the Standard. In the ejector the tracer gas passes through a critical orifice, entrains air from slots in the side of the ejector shell, and is distributed in the hood thru a wire mesh outlet. This ejector is positioned on the work surface inside a typical bench hood 6 in. behind the hood sash and in each of three lateral test positions, which are 12 in. from the left inside wall, center of the hood and 12 in. from the right inside wall. The ultimate test result is the maximum reading, of the three averaged values, of the three test positions. For walk-in hoods it is not representative of hood usage to place the ejector on the hood floor. The Standard does not define a height for the ejector; however, I use 18 in. up from the hood floor.

The test gas flow rate from the ejector is a critical factor in determining the hood performance rating. The ejector should be calibrated at the flow rate to be used for the test no more than 24 hours prior to the commencement of testing, immediately after testing, and anytime the critical orifice is changed. ASHRAE 110-1985 calls for a bubble flow meter to be used in calibration. I have found that an aluminum plate with a short pipe stub can be fastened to the critical orifice plate discharge hole using an "O" ring seal. This is then connected with rubber or plastic tubing to a Mylar bag that has been deflated and placed in a box that is 4 liters in volume. Once a 4-lpm flow has been established, you note the pressure on the gage and use this for your flow rate calibration. This box can be used for the calibration of 1-, 4-, or 8-lpm flow rates by either extending or reducing the time of inflation.

To conduct the test you turn on the tracer gas at the prescribed pressure and take a reading on the detection instrument every 10 seconds. This can be the manual notation of a visual reading, the action of an electronic "data logger," or the result of a computer software program. The Standard calls for a 10-minute test duration. I have reduced this to 5 minutes since in all of my testing I have seen most hoods indicate poor performance in the first 3 minutes. A few have shown initial signs of failure in the 4th and 5th minutes but none after that time frame. The only time I use the 10-minute, or even longer, time base has been to investigate room air patterns and not to determine a pass/fail condition.

To report the results of the test we take the average reading in ppm for the test period and record it as:

$$xxAMyyy \quad or \quad xxAUyyy$$

where:

xx is the flow rate in lpm.

AM is "AS MANUFACTURED" and is usually performed by a manufacturer in a test room and with no apparatus located inside the hood.

AU is "AS USED" and is done in the resident laboratory with the equipment inside the hood as it was being used.

yyy is the sampled test gas concentration, in ppm, as determined by the test protocol.

As an example, a rating of 4.0 AU .01 would indicate that a hood tested AS USED in a laboratory, at a test gas flow rate of 4 lpm, had a control level of 0.01 ppm.

To ascertain whether a hood performs well, the ACGIH has suggested that a hood tested at 4 lpm under AM conditions should have a control level not to exceed 0.05 ppm. For hoods tested under AU conditions the control level was set at 0.10 ppm.

Now for a couple of tips to make testing under ASHRAE 110-1985 easier and hopefully accurate.

Sulfur hexafluoride, the most widely used test gas, comes in moderately high-pressure cylinders. The regulator/pressure reducer, which has the pressure gage on it, fits to the cylinder with a metal-to-metal mating of surfaces. This makes it difficult to achieve an absolute leakproof seal. I suggest that you purchase your SF_6 in a small enough cylinder so that this cylinder can be placed inside of an adjoining hood to the one being tested. Since the pressure-indicating gage and valve are now on the cylinder, there is no need to provide a second set in the system. The SF6 is at the reduced pressure as it leaves the gage connection, and therefore you can use rubber or plastic tubing to the ejector.

When you are conducting the ASHRAE protocol, take a room background reading at regular intervals during the testing sequence. You can deduct this background from your testing readings for more accurate results. If the background gets pretty high, you have a hood-to-room leakage or a reentry situation that should be addressed before continuing with the test procedure. The background should stabilize at some reading in the 0.01-ppm range.

How do you determine the release rate that you should use for your testing?

When water is boiled on a 500-watt hot plate, the vapor release rate is approximately 8 lpm. The amount of solvent vapor released while pouring acetone, chloroform, or toluene from one beaker to another is approximately 1 lpm. The 4-lpm rate is a happy medium between the two examples shown. I like 4 lpm as being representative of a greater percentage of actual laboratory procedures where you do pour some solvents and you do heat some water, but perhaps not boil the water. The challenge release rate is up to you. You will get more favorable test results at 1 lpm than at 4 lpm, but you may not have a safer hood.

One last, but major, decision you must make is in the selection of the detection instrumentation. There are currently two dominant systems: the Foxboro Miran 1A infrared spectrophotometer and the ITI electron capture chamber. Here is a little guidance to help you make a choice.

The Miran 1A is an excellent laboratory instrument and is capable of analyzing a broad spectrum of gases. Properly used and maintained it is stable, easily calibrated, and more than reliable. New, this instrument costs in the $15,000 range; used and rebuilt its cost falls somewhere between $8000 and $10,000. As an instrument it is not easy to transport on a commercial airliner. In its padded carrying case the Miran is bulky and heavy, and it is much too fragile to check with regular baggage; it must be carried on the plane. It does not fit in the overhead bin and therefore a sympathetic flight attendant must put it in a closet or behind the last row of first class seats. If you travel by air to test hoods think twice before you purchase the Miran 1A.

The ITI instrument is capable of detecting any electron capture gas such as Freon and SF_6. It is reasonably small and can be battery operated if required. On the first instruments it was not easy to get a good field calibration, and the instrument had a tendency to drift. This appears to have been corrected in the newer models, number 120 and above. From a cost standpoint a new one is in the $11,000 range, and there is no great market in used ones.

As with all instruments they both have their quirks. Both are sensitive to electrical line disturbances (electric drills, mixers, etc.), particularly on the most sensitive scale. You can correct this by seeking out a controlled circuit, or by using the ITI battery pack. Most facilities have these circuits for their computer equipment so they are not hard to find. The Miran 1A plots easily and directly on a single pen recorder. The ITI system does not, and you have to set up a "data logger" interface or a computer software system to normalize the rapid rise and fall of the readings to get a reasonable curve for an output. The Miran instrument shows a running average because of the large volume of the test cell.

Add up the various factors of each system and your costs are not too far apart. Remember you have to buy the ejector (approximately $1200), a used manikin ($100), a method of flow rate calibration, tubing and the SF_6, carts, recorders, maybe a data logger, and so on.

One last warning before we go any further. DO NOT buy into a system that only gives you a single weighted value for the test period. It is worthless, though less expensive. If you have any thoughts of knowing more than pass/fail (why is my hood not performing well), don't get boxed in. Soon we start to see what ASHRAE 110-1985 can show you in regard to hood and building performance. If you think you will save money by doing grab sampling, think again. Save your time and your money; do qualitative tests for now and, when you have enough money saved, buy the proper equipment.

"AS USED" AND "AS MANUFACTURED" TESTING

Each and every laboratory hood should eventually be tested in the "AS USED" mode. When you are buying a new hood be sure that the manufacturer tests a model identical to the one you have purchased in the "AS MANUFACTURED" mode. I do want to point out that there can be a very marked difference between the hood test facility used by the manufacturer and the laboratory in which you work. ASHRAE 110-1985 places no limits as to how perfect the manufacturer can establish the air currents, traffic patterns, and so on, and as a result the AS MANUFACTURED facility can be pretty ideal. One manufacturer has a test facility that excludes people from the room during the test protocol and has a disturbance-free air make-up system. Make the manufacturer furnish a drawing or sketch of the facility showing locations of the air diffusers, how many people are present in the room during the test, room cross-drafts, and so on.

In the generic specification in Chapter 10 you will note that I have added some challenge for the AS MANUFACTURED testing. I have included cardboard boxes and paint cans to simulate the effect of equipment. People are present and then there is a "walk-by" directly behind the manikin during the test. I feel that these are all necessary ingredients to an AS MANUFACTURED test. While they are not part of the Standard, there is nothing keeping you from adding them to your specification or purchase agreement. For heaven's sake don't just put in a one-sentence requirement that the hood be tested in accordance with ASHRAE Standard 110-1985. It may take your engineer or purchasing people 10 minutes longer, but you will get a hood that may be 1000 times safer.

"AS INSTALLED" TESTING

There is a void in the nomenclature for testing newly installed hoods. They are not either of the other two classifications. I propose that ASHRAE establish a new category to cover this and I suggest "AS INSTALLED." When you read Part 3.06 of the specifications outlined in Chapter 10, you will see that I have added some simulated apparatus and some in-lab performance challenge. I feel that these areas should be addressed for the AS MANUFACTURED and AS INSTALLED rating.

In the discussions that follow, regarding actual field testing of fume hoods, I have used the Miran 1A instrument. For your information, and to give you some degree of comfort with the results, I am going to list the principal settings used during the test.

1. Path length = 21.25 meters.
2. Wave length Freon-12 = 9.1 microns SF_6 = 10.6 microns.
3. Slit width = 1 mm.
4. Response time = 10 seconds.
5. Scale = 0.025 full scale.

The ASHRAE 110-1985 (or 110-1993) protocol can furnish you with test results that can be used to do routine pass/fail testing; to detect building air reentry problems or duct work leakage; to point out improper equipment location within the hood, unacceptable hood housekeeping, disruptive room air patterns, poor traffic patterns within the room, and so on down a very long list. The best part about using the complete protocol is that it can readily demonstrate to you when you have corrected one or all of the problems.

TYPICAL HOOD TESTS

This laboratory was 18 ft by 22 ft and housed four each, 5 ft air foil styled fume hoods. The hoods were located away from the corners of the room with one on each side wall of the room and two mounted back to back on a center island. The blade type air entry devices were randomly located in relation to the hoods and had discharge velocities that ranged from 150 to 250 fpm. Both the facility and hood housekeeping were good with only a moderate amount of equipment placed in each hood. The hoods had a vertical rising sash and one horizontal sliding plastic safety shield 16 in. wide.

The first test had the safety shield in place on the right side of the hood sash area. The face velocity was evenly distributed with an averaged value of 88 fpm. The test was conducted at the center position of the remaining 36-in. sash opening. The results are shown in Figure 9.1.

The test was aborted after 1 minute since the test indication was off-scale starting at 15 seconds, and continuing for an additional 15 seconds which indicated that it was not a random peak that would diminish with time.

For the second part of the test the safety shield was removed from the sash opening; the face velocity decreased to 50 fpm; the test was conducted at the center of the full sash opening. The results as shown in Figure 9.2.

The hood rating for this test was 4.0 AU .003.

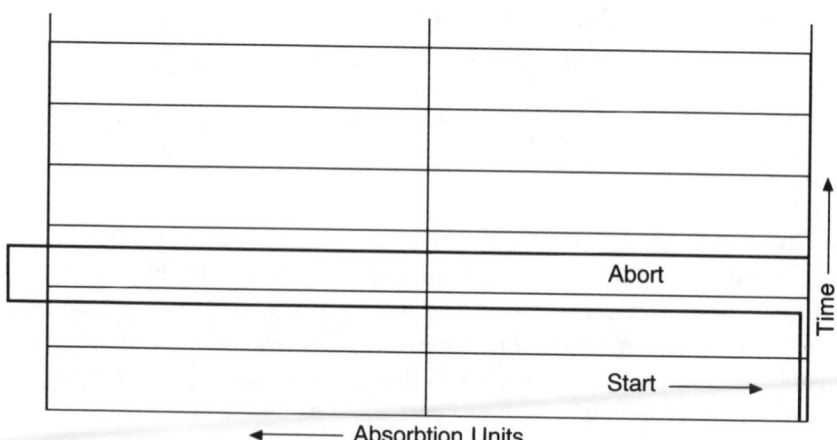

FIGURE 9.1 *ASHRAE 110-1985 test results of a malfunctioning hood.*

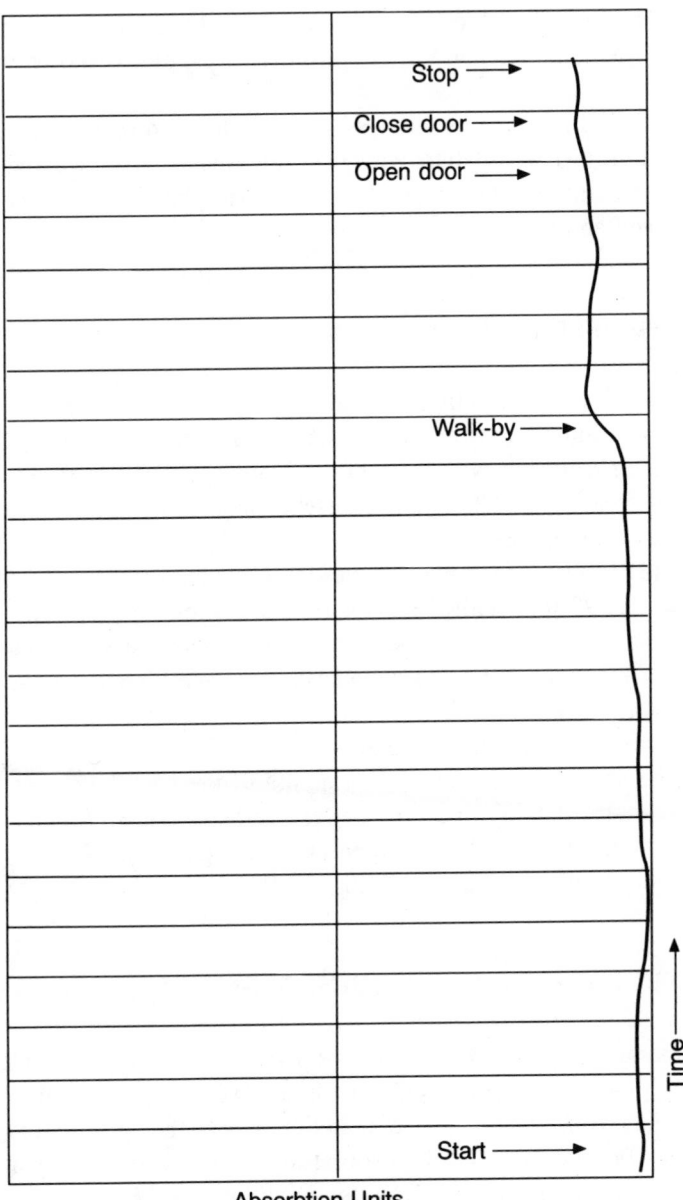

FIGURE 9.2 *Retest of hood with safety shield removed and face velocity at 50 fpm.*

To determine the effect of the safety shield track on the hood performance it was removed and the test rerun. The test plot is shown in Figure 9.3.

The hood rating for this test was 4.0 AU .001. You will note that near the end of the test period there were two times that a person walked directly behind the manikin (shown as "walk-bys") and the main laboratory door was opened and closed.

BACK BAFFLE SLOT ADJUSTMENT

This laboratory was 10 ft by 22 ft and had two each, 5 ft, constant-volume, square faced fume hoods in the room; they had vertical rising sashes. The make-up air diffusers were a standard blade type unit with an average discharge velocity of 200 fpm. Housekeeping was good, with minimal apparatus in the hoods.

For the first test the top back baffle slot was 2 in. open. The average face velocity was 95 fpm with the velocity distribution higher at the top and center than at the bottom of the sash opening. The test was conducted at the center location of the full open sash area. See Figure 9.4 for the test results.

The hood had a rating of 4.0 AU .370.

For the next part of the test the top slot of the back baffle was closed down to a $\frac{1}{2}$-in. opening; the average face velocity was 85 fpm and evenly distributed across the full sash opening. See Figure 9.5.

The hood rating improved to 4.0 AU .04.

TRAFFIC PATTERNS AND DOOR OPENINGS/CLOSINGS

This laboratory was 12 ft by 18 ft with two each, 5-ft air foil styled fume hoods with vertical rising sashes The hoods were located along each side wall of the room. There was one main doorway to the corridor and a small secondary door to the adjoining laboratory. People were moving about the room during the test period. Both the laboratory and hood housekeeping were very good.

The first test was performed with the main hallway door secured; personnel entered the room through the secondary door to the adjoining

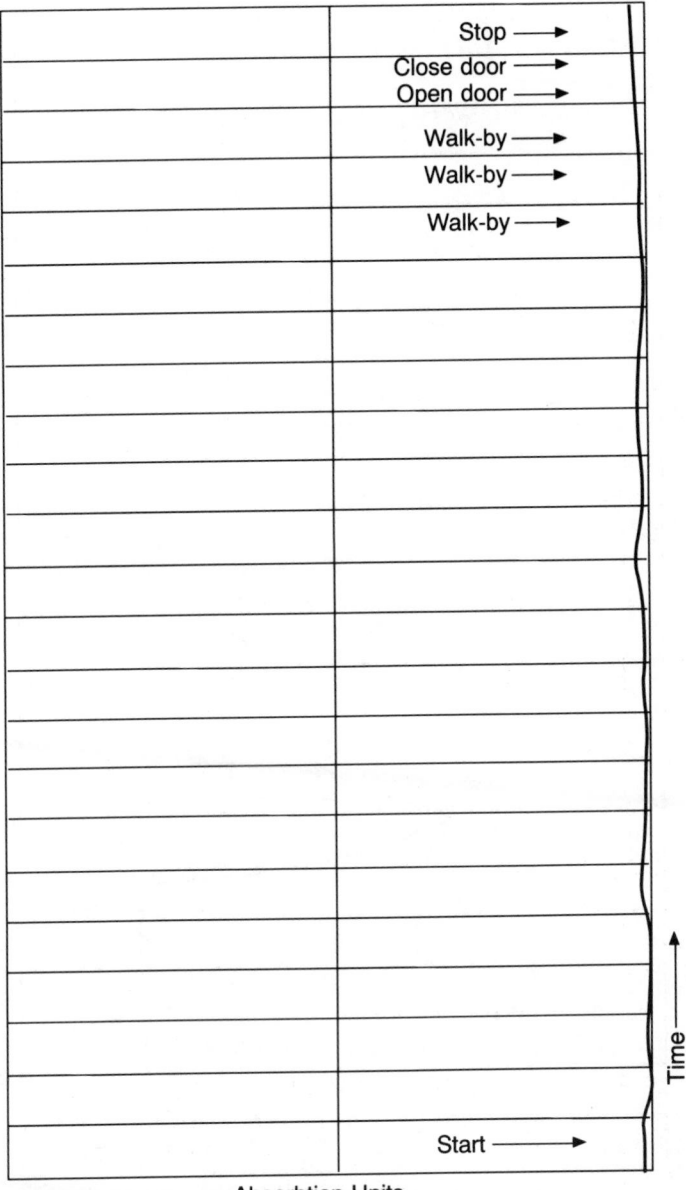

FIGURE 9.3 *Test results with safety shield track removed.*

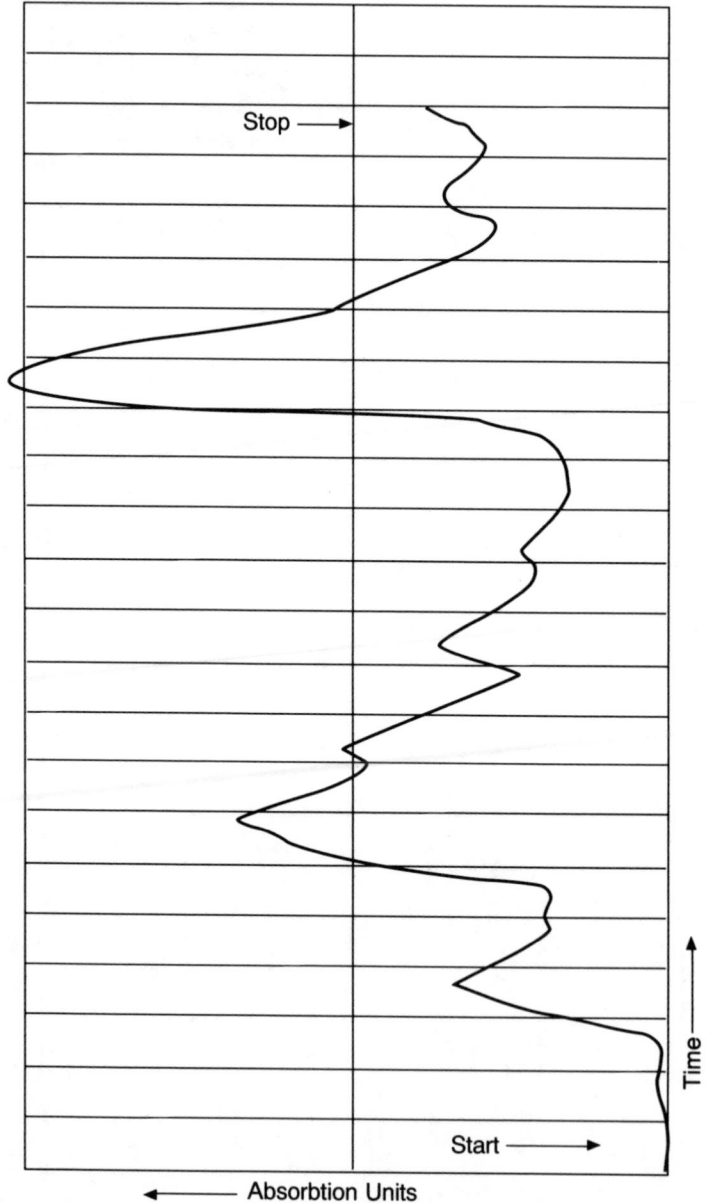

FIGURE 9.4   *ASHRAE 110-1985 test results with back top baffle-slot fully open.*

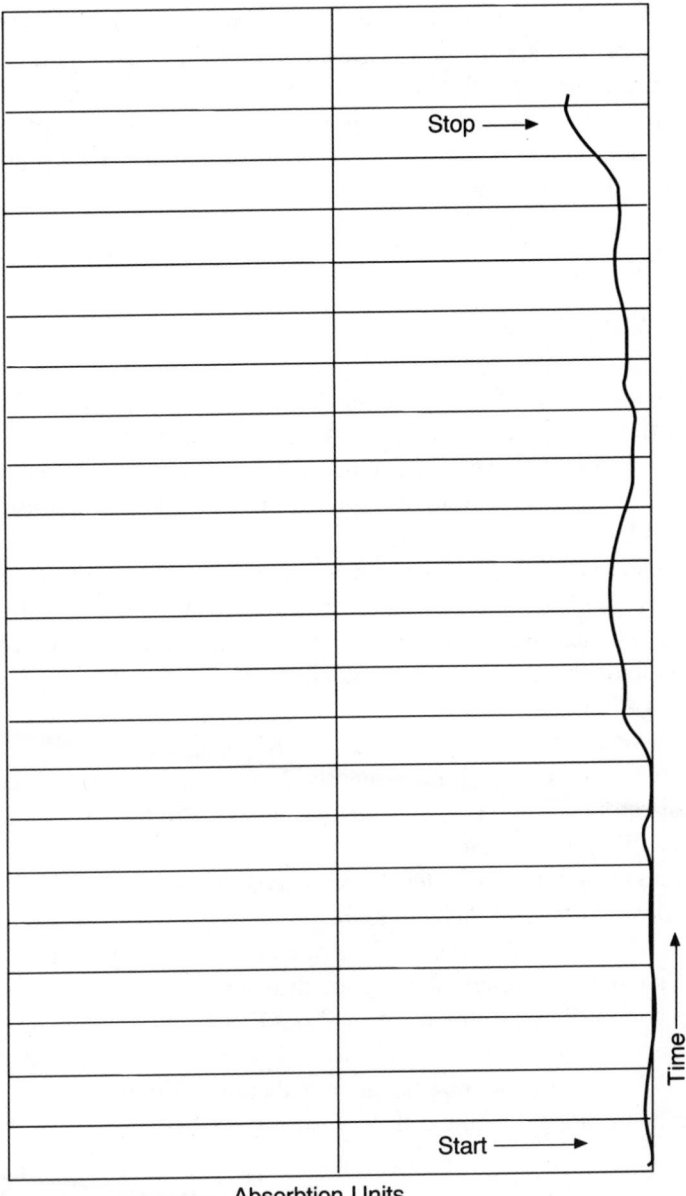

FIGURE 9.5 *Retest with back baffle closed to a $\frac{1}{2}$-in. opening.*

room. Face velocity of the hood was 123 fpm and the test was conducted at the center position of the full sash opening. Refer to Figure 9.6.

The hood rating for this test sequence was 4.0 AU .001.

The test was repeated with access to the room by way of the main corridor doorway. The large spike shown in Figure 9.7 resulted from opening and closing the hallway door.

The hood rating for this test was 4.0 AU .159.

ROOM AIR PATTERNS

This laboratory was 18 ft by 22 ft and had six each 5-ft air foil styled hoods with vertical rising sashes and a 16-in. horizontal plastic safety shield. The make-up air diffusers were of a standard blade design with terminal velocities that ranged from 200 to 450 fpm. Both hood and room housekeeping were good.

The hood selected for testing was located on the center island of the laboratory and had the safety shield positioned on the left side of the sash opening. The test was conducted in the center position of the remaining 36-in. sash opening and the face velocity was 130 fpm. The test results are shown in Figure 9.8.

The hood rating for this test was 4.0 AU .14.

For the second test of the same hood an 18-in. by 24-in. rectangular piece of cardboard was suspended at a right angle to the closest ceiling air entrance diffuser. See Figure 9.9.

The hood rating for this test was 4.0 AU .49

For the next test, in the series, the piece of cardboard was changed to a 75 degree angle and a second piece of cardboard was suspended at a 180 degree angle to a second air entrance diffuser. See Figure 9.10.

The hood rating for this test was 4.0 AU .01.

In the final test in this series, the second piece of cardboard was changed from 180° to 120°. The test results are shown in Figure 9.11.

The hood rating for this test was 4.0 AU .001.

DUCTING LEAKAGE

This was a large and diverse laboratory 22 ft by 30 ft that contained eight fume hoods of varying sizes including two walk-in hoods. The make-up

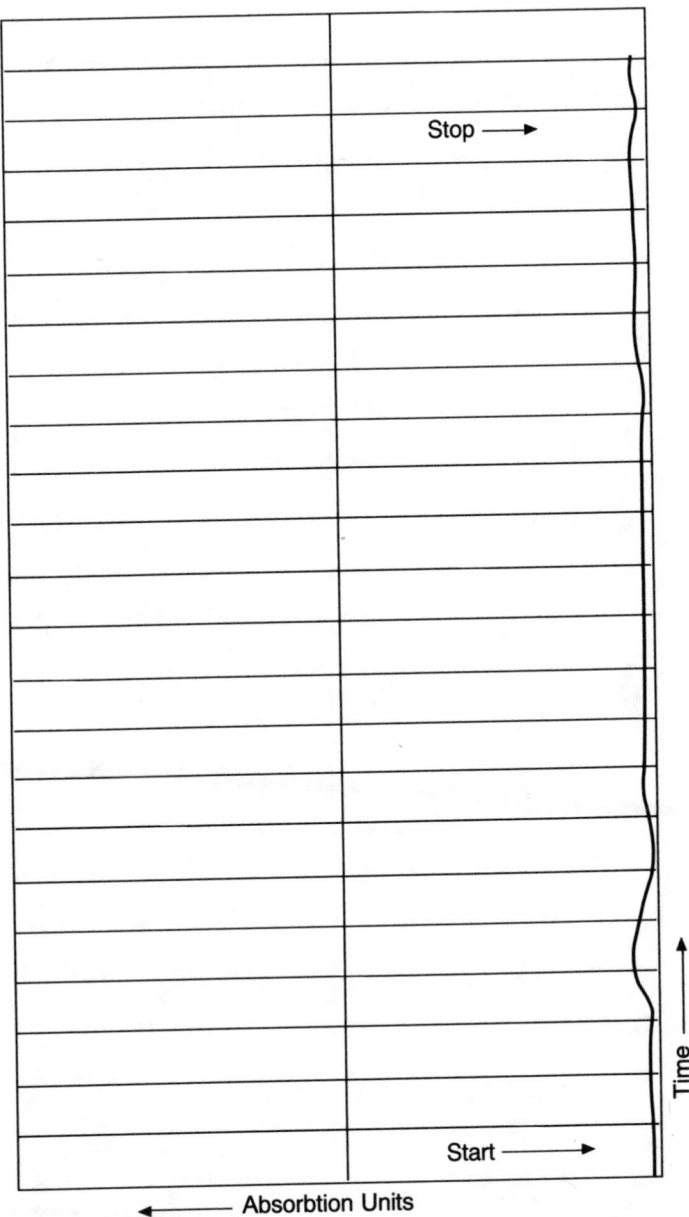

FIGURE 9.6 *ASHRAE 110-1985 test results with restricted room access.*

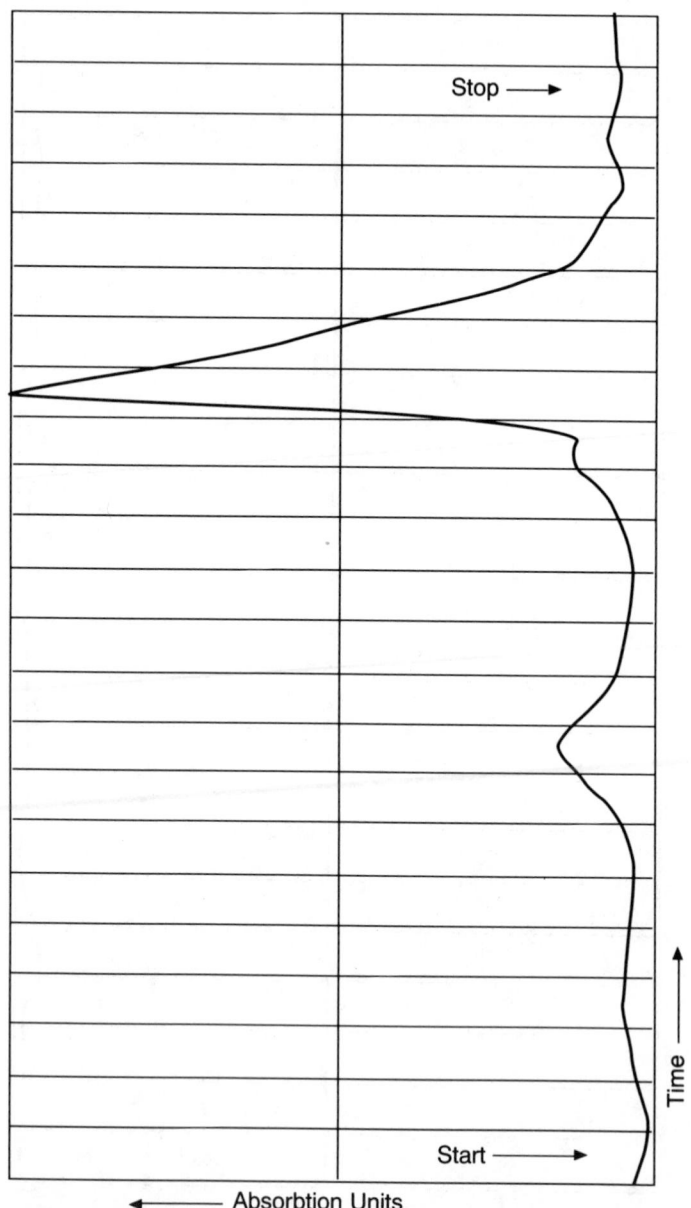

FIGURE 9.7 *Retest with hallway door access to laboratory.*

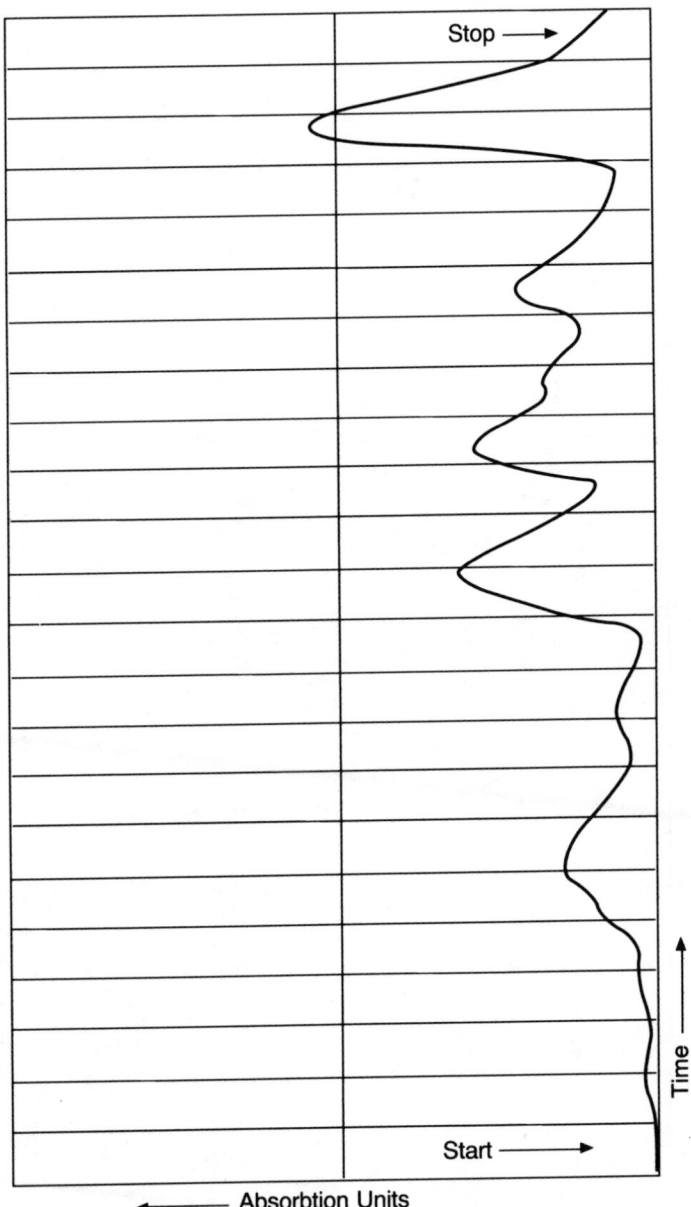

FIGURE 9.8 *ASHRAE 110-1985 test results of a laboratory with air entry prob-lems.*

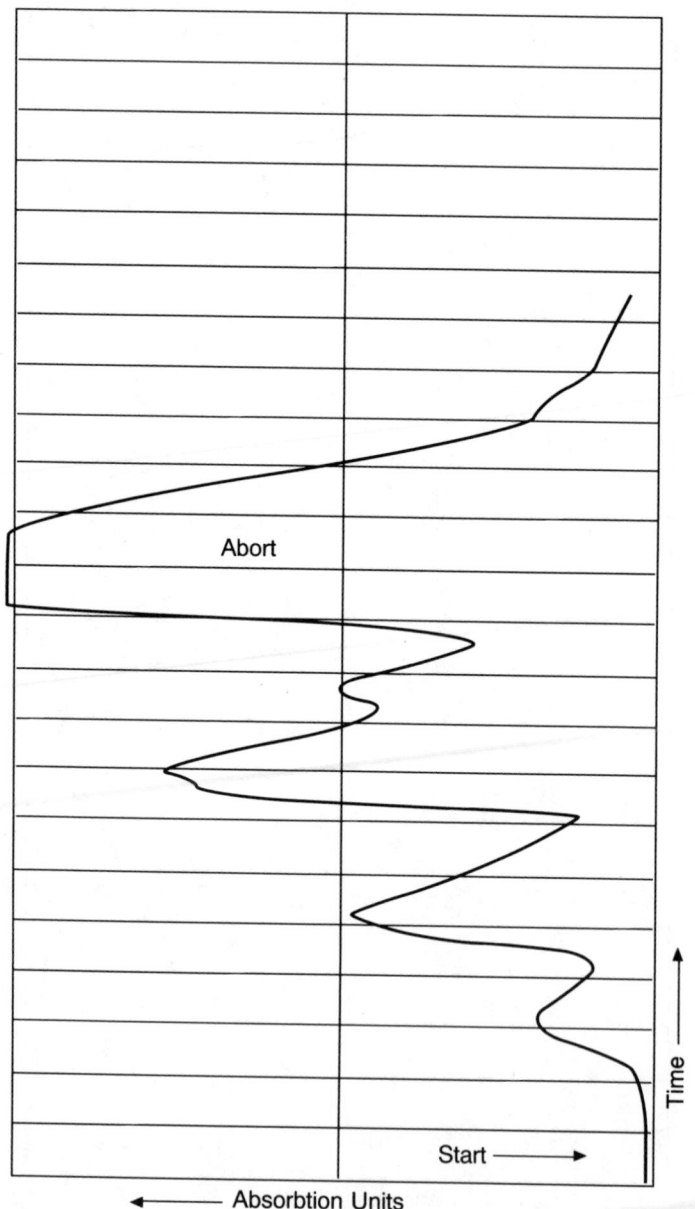

FIGURE 9.9 Test results with cardboard suspended at right angle to ceiling air entrance diffuser.

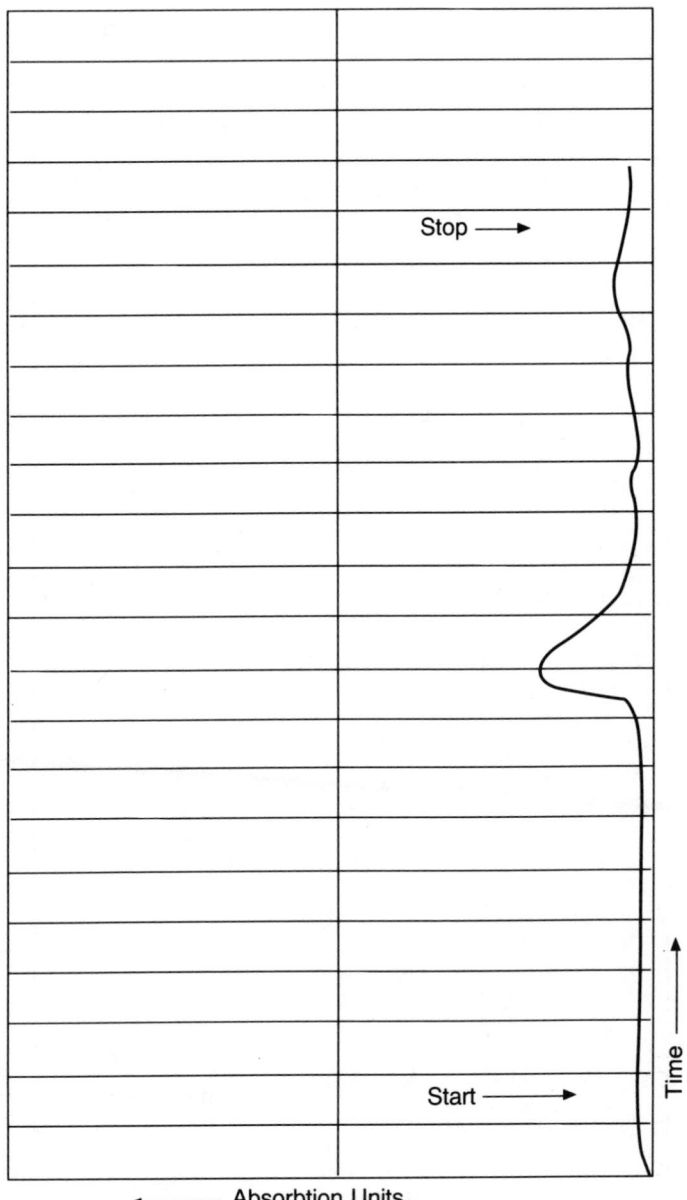

FIGURE 9.10 Test results with two pieces of cardboard suspended at 75° and 180° angles to two different air entrance diffusers.

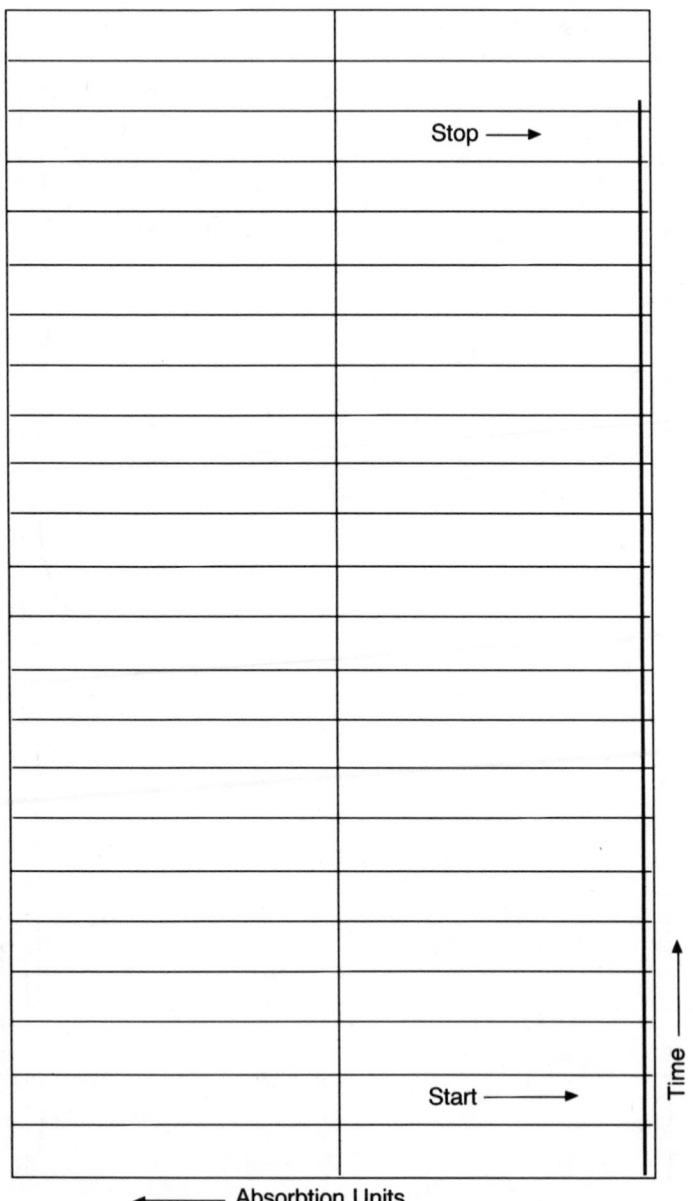

FIGURE 9.11 *Test results with cardboard suspended at 75° and 120° angles to two different air entrance diffusers.*

air system used standard blade diffusers with an average discharge velocity of 150 fpm. Both laboratory and hood housekeeping were good.

When work involving a solvent that had a characteristic odor was done in a specific 6-ft air foil styled hood, the odor of the solvent was quite noticeable some 15 ft away at the entrance of the room. A test of the hood showed excellent containment. A probe sampling of the air make-up diffuser in the entrance area of the laboratory did not show any SF_6 entrainment.

A ceiling tile was opened above the entrance area of the room. Sulfur hexafluoride was discharged into the hood at a rate of 8 lpm and a timed sequence probe sampling at the open tile area indicated SF_6 entrainment starting at 3 minutes. Refer to Figure 9.12.

Sulfur hexafluoride was then discharged into other hoods in the room and SF_6 contamination was not detected.

Several large ceiling tiles were removed to allow an inspection of the exhaust ducting of the 6-ft hood under investigation. While SF_6 was discharged into the hood, a probe sampling was made around the exhaust ducting and some areas showed SF_6 contamination. A close inspection of the ducting showed the major causes of the outward air leakage.

The plenum area, above the ceiling was shallow (approximately 3 ft) and was well occupied with ducts, electrical conduits and structural beams. The ducting was of welded stainless steel construction and except for the hood duct collar it was all rectangular in design.

The hood duct collar (round) had been inserted into the rectangular exhaust ducting system but had never been welded. It was in this open joint that the SF_6 leakage was detected. While the exhaust ducting was under a negative pressure of approximately $1\frac{1}{2}$-in. of water gage (wg), there was sufficient turbulence at the open joint to allow outward leakage.

EXHAUST AIR REENTRY

This laboratory was 8 ft by 15 ft and had a single 4-ft auxiliary air, air foil style hood. Both hood and laboratory housekeeping were excellent.

The first test was conducted with the sash full open, center position for the manikin, a face velocity of 96 fpm and the auxiliary air volume at 50 percent of the exhaust volume. At 30 seconds the test was terminated since the concentrations of the test gas rose rapidly and did not retreat after 45 seconds. See Figure 9.13.

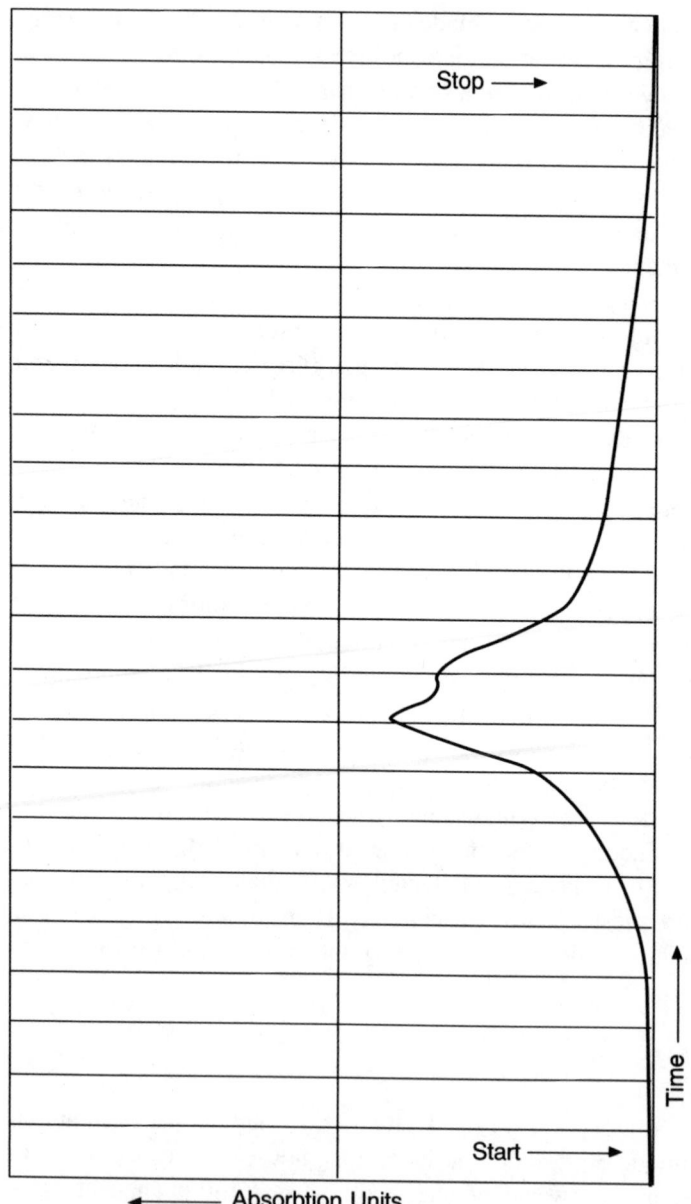

FIGURE 9.12 *ASHRAE 110-1985 leak testing.*

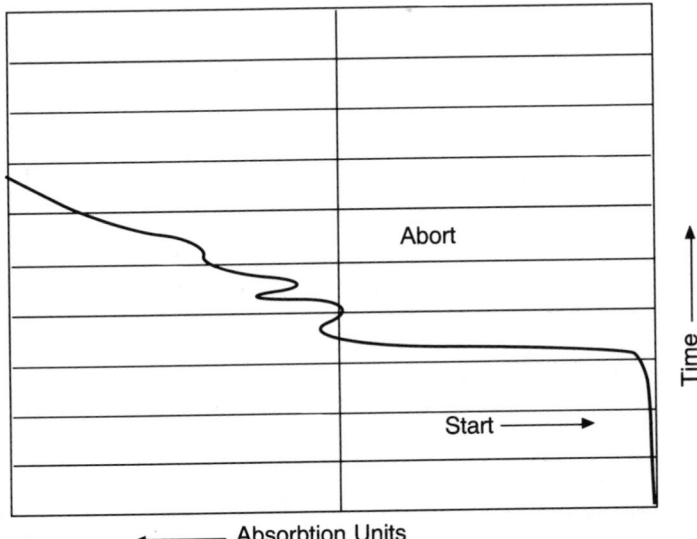

FIGURE 9.13   ASHRAE 110-1985 exhaust air reentry test: sash full open.

The test was repeated with the sash closed to 18 in. The face velocity increased to 180 fpm due to the sash height change. After 15 seconds the test gas concentration rose rapidly to 1.0 ppm, fluctuated in that range for over a minute, and then went off-scale. The test was terminated at $1\frac{1}{2}$ minutes. Refer to Figure 9.14.

A retest with the sash lowered to 15 in. gave much the same result.

At this point in the testing it became obvious that the test gas detection did not come from an outflow from the hood itself since if this had been the case the test gas concentrations would have decreased markedly as the sash was lowered.

As a final test the collection probe was placed in the auxiliary air plenum. In less than 15 seconds the instrument reading was off-scale. See Figure 9.15.

I did not have the opportunity to inspect the auxiliary air and exhaust air systems. I was informed that there was one fan for each and that this hood was totally independent from the building exhaust system. I would hazard a guess that the exhaust air blower and the auxiliary air blower were very close to one another. When the air was discharged on the roof area by the exhaust blower some was immediately entrained by the auxiliary air

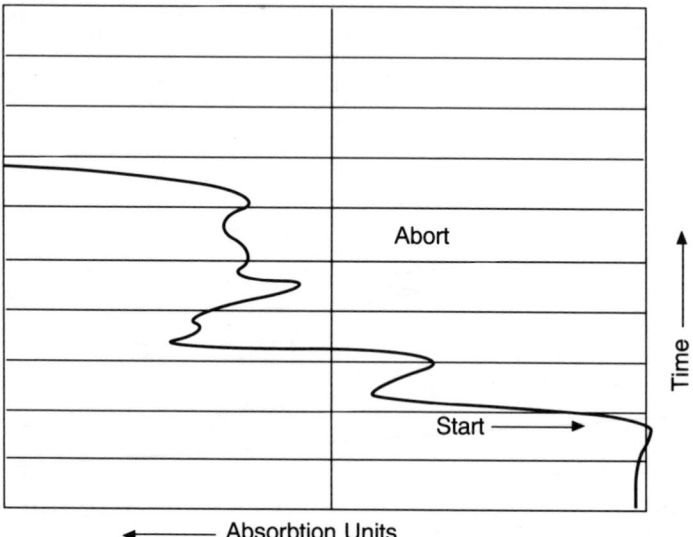

FIGURE 9.14 *Exhaust air reentry test: sash closed to 18 in.*

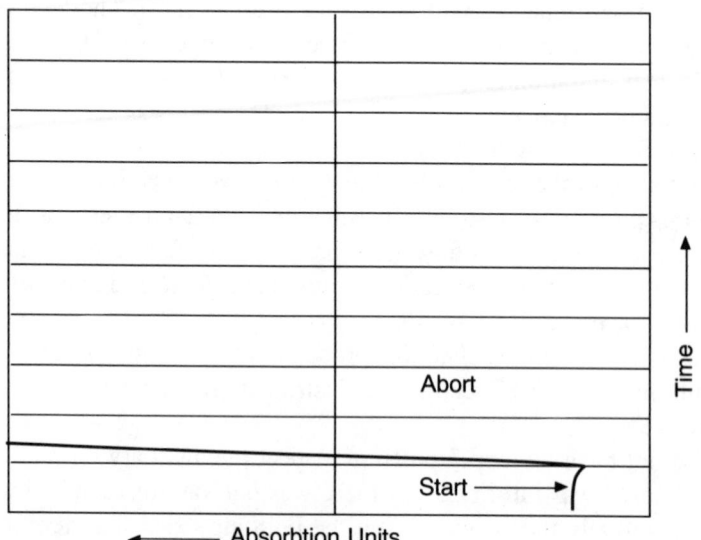

FIGURE 9.15 *Exhaust air reentry test: probe placed in auxiliary air plenum.*

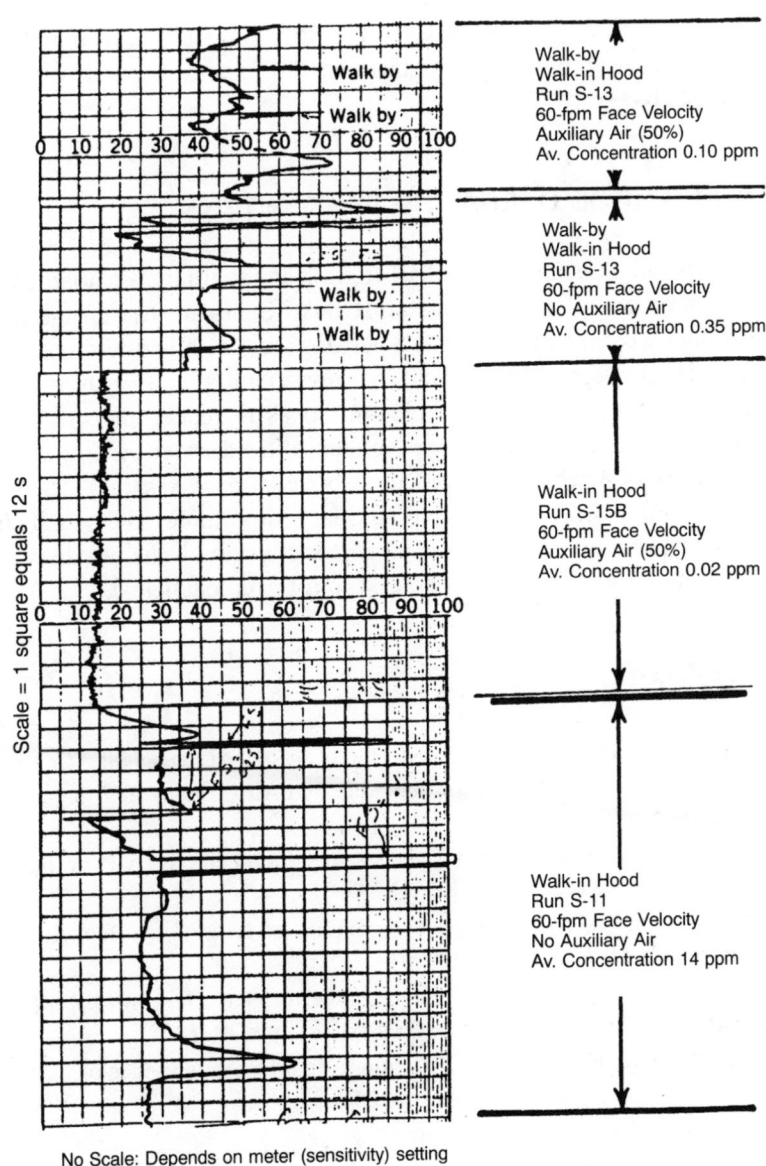

Walk-by
Walk-in Hood
Run S-13
60-fpm Face Velocity
Auxiliary Air (50%)
Av. Concentration 0.10 ppm

Walk-by
Walk-in Hood
Run S-13
60-fpm Face Velocity
No Auxiliary Air
Av. Concentration 0.35 ppm

Walk-in Hood
Run S-15B
60-fpm Face Velocity
Auxiliary Air (50%)
Av. Concentration 0.02 ppm

Walk-in Hood
Run S-11
60-fpm Face Velocity
No Auxiliary Air
Av. Concentration 14 ppm

Scale = 1 square equals 12 s

No Scale: Depends on meter (sensitivity) setting

10 ft Walk-in Hood—Horz. Sliding Sash
60-fpm Face Velocity
With and Without Auxiliary Air

FIGURE 9.16 *ASHRAE 110-1985 in front of test of prototype hood: performance with auxiliary air on/off and with/without room traffic challenge.*

system and returned to the laboratory. The solution could be as simple as moving one blower or adding a stack on the exhaust blower, or both correction steps.

CHANGING HOOD OPERATING PARAMETERS

I had designed a special purpose, 10-ft walk-in, auxiliary air hood for a major research complex in the eastern United States. In testing the prototype model we examined the hood performance with and without auxiliary air. The hood had three each, 3-ft horizontal sliding sash in three tracks, a very large auxiliary air plenum so as to reduce the discharge velocity of the auxiliary air, and a compensating damper system so the auxiliary air only discharged in the area above an open sash(es). The test was run at 60 fpm with a 50 percent auxiliary air volume. The composite chart shown in Figure 9.16 shows the performance with the auxiliary air on and off and with and without some room traffic challenge.

REFERENCES

1. Method of Testing Laboratory Fume Hoods. ASHRAE Standard 110-1985, American Society of Heating, Refrigerating and Air Conditioning Engineers. Atlanta, 1985.
2. K. J. Caplan and G. W. Knutson, "Laboratory Fume Hoods, A Performance Test," RP-70, *ASHRAE Trans.*, 84, (I)(1978).
3. F. H. Fuller and A. W. Etchells, "The Rating of Laboratory Hood Performance," *ASHRAE J.*, 21, 49–53(1979).
4. "Development of Quantitative Containment Performance Tests for Laboratory Fume Hoods," EPA Contract No. 68-01-6197 (June 29, 1982).
5. *Industrial Ventilation Manual*, 17th edition, American Conference of Governmental Industrial Hygienists, Cincinnati, 1982.

10

Specifications

I recently read an article editorializing on various magazines and it mentioned that *The New Yorker* was the most unread magazine published. If this is the case, then specifications for laboratory fume hoods must be the most unread section of laboratory construction specifications. Many a time have I pointed out to the architect and an errant manufacturer that what was delivered was not what was called for and all I got were some very blank stares from both. Then the usual, "Golly, I didn't see that." You can have only one answer: "Fix it"—regardless of the time delay and mess it may cause. Don't you worry about who pays for it: do not pay either the architect or the contractor their regular fees until it is fixed. Remember, you told them what you wanted, what you expected, and by heaven you are going to get it. After the building project is finished you are going to be there for a long time and the architect and the manufacturer are going to be long gone. I was associated with a major project where the director of research told the contractor to redo the ventilation system; it took over a month and delayed the building occupancy, but the decision was never regretted and the delay has long since been forgotten.

Now for specifications. It could be that one reason the hood section is lightly read is that it gets too long, ambiguous, and many times merely says what some manufacturer has told the architect will guarantee good hoods (as long as that manufacturer gets the contract). Some other company gets the contract for the hoods and you get a mess. I am for a hood specification that (1) does not favor one manufacturer, (2) spells out what I want, (3) lists the manufacturers that I will accept (in general terms, not specific product), and (4) spells out the basis for the acceptance of the installed product. Get involved in the writing of the specification, be sure everyone in the architectural office understands what you are saying and then stick with your guns, and the specifications, until you move into the facility. Don't be unreasonable, but be firm. Remember, it is your money, or ours if it is a government-funded project.

SECTION ____ LABORATORY FUME HOODS

Part 1.00 General
   1.01 Description
     A Work Included:
       1. This section includes, but is not limited to, all materials, labor, and equipment to furnish, deliver, and install all laboratory fume hoods, hood work surfaces, hood base cabinets (if included in contract), pipe chase for hoods, miscellaneous hood fillers, panels, and scribes for a finished installation.
       2. Furnish and deliver to the mechanical contractor, loose and in boxes, for his installation, all of the service cocks, bibs, faucets, sink bowls, cup sinks, lock nuts, and strainers for the laboratory fume hoods. The fume hoods will be prewired by the hood manufacturer (see Part 3.00 this section).
     B Related Work Elsewhere:
       1. Vinyl base molding
       2. Plumbing
       3. Electrical
       4. HVAC
   1.02 Quality Assurance
     A The manufacturer shall, from 1 year of date of completed installation and acceptance by owner, warrant all manufactured parts against defects in workmanship and materials. Any such parts, under normal use, that prove to be defective will be repaired or replaced without charge to the owner.
   1.03 Submittals

A Shop Drawings
1. Identify the location of all fume hoods in the project on a floor plan format.
2. Detail the laboratory fume hoods in related and dimensional detail with sections as required. Details to include notation of all specified items.
3. Locations for rough-in of plumbing and electrical services, sinks, valves, bibs, faucets, etc. as related to the fume hoods.

B Certificates
1. The fume hood manufacturer, shall include with their bid independent test results, from a nationally recognized testing laboratory, showing compliance with the finish specifications for the fume hood painted parts.

1.04 Delivery, Storage, and Handling

A The laboratory fume hoods and associated materials shall be shipped, stored, and handled to prevent damage to the product.

B All work surfaces will be specially protected from damage during transit, storage, and after installation.

2.00 Product, General

2.01 Laboratory Fume Hoods

A The laboratory fume hoods will be similar to those manufactured by one of the following companies:
1. Duralab Equipment Corporation
2. Fisher Scientific Company
3. Hamilton Industries
4. Kewaunee Scientific Corporation
5. St. Charles Manufacturing Co.

B These companies may use their standard manufacturing detail; however, they must conform with the specific details of Part 3.00 of this Section.

2.02 Steel Panel Finish

A After fabrication each shell panel, filler, scribe, etc, will be chemically cleaned, treated, and dried prior to paint application. A prime and finish coat will be applied and baked as per each manufacturers's process with a final coating thickness of from 1.75 to 2.25 mils.

B Chemical Resistance
Chemical spot testing shall be conducted by applying 5 drops of the reagents listed below to a marked area on the painted steel

surface. On reagents 1 through 9 a 65-mm watch glass shall be placed convex side down against the surface. On reagents 10 through 18, a 65-mm watch glass will be placed convex side up over the reagent. The test will continue for 1 hour, at which time the watch glasses will be removed and the surface washed with a mild detergent and dried. After 1 hour the designated surfaces will be inspected and graded as to:

1. NE = no effect
2. LG = loss of gloss
3. SB = slight blistering
4. BM = bare metal

Test results that show slight blistering or bare metal will not be acceptable.

1. Acetic acid (98%)
2. Hydrochloric acid (37%)
3. Nitric acid (35%)
4. Nitric acid (10%)
5. Phosphoric acid (75%)
6. Sulfuric acid (60%)
7. Sulfuric acid (25%)
8. Ammonium hydroxide (28%)
9. Sodium hydroxide (25%)
10. Acetone
11. Carbon tetrachloride
12. Ethyl acetate
13. Ethyl alcohol
14. Ethyl ether
15. Formaldehyde (37%)
16. Methylethyl ketone
17. Phenol
18. Xylene

Hardness: The paint test samples shall exhibit a minimum hardness of 4-H using the pencil hardness test procedure.

2.03 Service Fittings and Fixtures

A All laboratory service fittings and fixtures shall be as manufactured by the Water Saver Faucet Company or an approved equal. Fixtures, including handles, shall be color coded to indicate the proper service. Refer Part 3.03.A.13, this section.

3.00 Fume Hood Details

3.01 Related Work

  A Refer to Part 1.01, this section.

  B The fume hood manufacturer will supply and prewire (in accordance with the National Electrical Code) all electrical fixtures, switches, lights, outlets, etc, that are directly related to and are part of the fume hood superstructure. This wiring will terminate in a junction box on the top, back right side section of the superstructure. The electrical contractor will be provided with a schematic of this wiring, and it shall be part of the hood manufacturer's submission drawings. This drawing will be approved by the architect. The number, type, and location of electrical fixtures will be as shown on the drawings.

3.02 Warranty The fume hoods will carry the warranty as outlined in Part 1.02 of this section. They shall meet the standard of quality as established by the manufacturer's specifications that were in effect at the time of bid, by accepted practices of the fume hood industry, and those added details outlined in Part 3.03 of this section.

3.03 Fume Hood Design and Construction

  A The following details will be made part and parcel of the design and construction of the fume hoods:

    1. The fume hood lining material shall be $\frac{1}{4}$ in. thick, non-asbestos cement board with physical and chemical properties equal to Flexboard II as manufactured by the John Mansville Co. Stainless steel lined hoods, where indicated on the drawings, shall be as per Parts 3.04 and 3.05 of this section.

THIS IS WHERE YOU SHOULD INSERT THE SPECIFICATIONS OR NOMENCLATURE OF A LINING MATERIAL OTHER THAN CEMENT BOARD OR STAINLESS STEEL IF THESE MATERIALS ARE NOT BEING USED. REFER TO MANUFACTURERS' CURRENT CATALOGS FOR MATERIAL DESCRIPTIONS.

    2. The cement board liner will be finished on all surfaces with an air-dry, white, dull or semigloss finish with excellent chemical resistance that can be touched up in the field. The hood manufacturer will submit a test report for this finish covering the chemicals outlined in Part 2.04 of this section.

3. The hood work surfaces will be molded epoxy resin $1\frac{1}{4}$-in. thick with a $\frac{3}{8}$-in. indent on all four sides. There will be a ledge, back of the rear indent, so as to be able to pass the $1\frac{1}{2}$-in. PVC vents through the work surface from either solvent or acid storage cabinets at this larger dimension. The front ledge will extend only to the back edge of the bottom front air foil. Where shown on the drawings, these work surfaces will be type 304 or type 316 stainless steel.

4. The bottom front air foil will be a rigid formed cross section of 16-gage mild steel that will be coated with Kynar (Pennwalt Corporation, Philadelphia, PA) or pinhole-free Teflon. It will have a 2-in. wide back section that acts as a stop for the sash and as a straightening section for the air passing under the foil. This flat section will slope down at an angle of from 20° to 30° to the front plane of the hood. There will be a 1- to $1\frac{1}{4}$-in. space between the work surface and the bottom side of the air foil.

5. The sash track will be Kynar coated steel or an acceptable plastic material.

6. The sash frame will be type 316 stainless steel with a #4 satin finish.

7. All metal parts exposed to the interior of the hood chamber, except the duct collar, will be coated with Kynar or pinhole free Teflon or be fabricated from an approved plastic material. Finishes must be submitted to the owner for testing prior to fabrication.

8. Duct collars will be type 316 stainless steel, mounted from the interior of the hood chamber with appropriate fasteners and will have the following dimensions:
   a. 5-ft hood = 12 in.
   b. 6-ft hood = 12 in.
   c. 8-ft hood = two each at 10 in.
   d. 10-ft hood = two each at 12 in.

9. These hoods will be of an aerodynamically designed entrance shape. The side wall thickness can range from 4 to 6 in.

10. The back baffle will have a fixed top slot $\frac{1}{2}$ in. wide, a fixed center slot $1\frac{1}{2}$ in. wide located 14 in. up from the work surface and a fixed $2\frac{1}{2}$ in. bottom slot. The back baffle shall

have a 2 in. clear spacing to the back surface of the hood superstructure. The top section of the baffle will act as a plenum that will extend at least $1\frac{1}{2}$ in. beyond the front of the duct collar opening.

11. Sash leakage, when the sash is open, shall be minimized by installing a Teflon wiper blade assembly between the back of the sash and the hood superstructure. The wiper assembly shall not impede the sash operation.

12. The sash weights for bench hoods 6 ft and smaller should be in the side walls of the hoods. The sash weights for hoods over 6 ft may go down behind the superstructure. Rear-mounted weights must be capable of replacement (if required in the field) without disconnecting service fittings or moving the superstructure.

13. All hood side wall valves will be of a front-mounted design with direct access to the valve stem and seat. The valve handles, interior bibs, and/or spouts will be appropriately color coded, according to service, using a Water Saver Faucet Company "labcote" finish, or an approved equal. The location, number, and identity of valves will be as shown on the drawings.

14. Cup sinks will be molded epoxy resin, 3 in. by 6 in. oval. Their location will be as shown on the drawings.

15. Vertical and horizontal cement board panels will be fastened one to another, as required, with an exterior framing system or by the use of multiple nonmetallic blocks that will be a minimum of 1 in. by 1 in. by 3 in. and spaced no more than 6 in. apart.

16. The lighting will be external relamping, two tube fluorescent fixtures on the maximum length(s) available for the fume hood width. Fluorescent tubes (daylight) will be furnished, but packed separately, by the fume hood manufacturer and will be installed by the hood manufacturer on the job site. There will be a vapor tight, $\frac{1}{4}$-in. gasketed, safety glass panel between the light fixture and the hood chamber.

17. There will be an interior access opening in each side wall of a 12-in. by 18-in. minimum size.

18. The sash, when fully raised, will provide an opening of at

least 31 in. from the hood work surface to the bottom of the sash.

IF YOU ARE GOING TO HAVE A COMBINATION HORIZONTAL/ VERTICAL OR A HORIZONTAL SASH IT SHOULD BE INSERTED HERE. YOU WILL HAVE TO REDO 3.03.A.5, 3.03.A.6, 3.03.A.11, AND 3.03.A.18 AS REQUIRED.

19. The acid and solvent storage cabinets, if they are located under the hood superstructure, will be vented through the work surface with $1\frac{1}{2}$-in. PVC piping. The pipe will extend 2 in. above the work surface and be sealed to the work surface with black "smooth-on." The solvent storage cabinets will have a flame arrestor screen located on the inside of the solvent storage cabinet. The hood manufacturer is responsible for furnishing and installing the venting system including all materials.

3.04 Perchloric Acid Fume Hoods

1. The hood liner shall be cove-welded, type 316 stainless steel.
2. The construction details shall follow the manufacturers' general standards for this type of hood. It shall, however, conform to the details as outlined in Part 3.03 (this section) as applicable.
3. The external sash frame will be as per 3.03.A.5. The internal section of the sash frame and sash guides will be rigid machined PVC.
4. The spray-down pipe, located at the top of the superstructure and behind the baffle, will be $\frac{1}{2}$-in. PVC or 316 stainless steel pipe with four rows of $\frac{1}{16}$-in. holes drilled on 1-in. centers. The resulting spray pattern will thoroughly cleanse the area behind the baffle. No splashing of water over the baffle and onto the hood work surface will be permitted. This washdown system will be prepiped from the valve to the pipe proper by the hood manufacturer. The hole pattern will be in proper alignment.
5. The light fixture will be (incandescent) explosion-proof construction, bulb included (150 watt). The wiring to the switch does not have to meet NEC code for class rated wiring.
6. There will be no interior access openings in the hood side

walls. There will be an external access opening in the hood shell for each side wall.

7. Internal hose bibs and spouts will be molded PVC.

8. The duct collar will be welded into the hood liner. The collar will extend $\frac{1}{8}$ to $\frac{1}{4}$ in. inside the liner roof and will be welded on the interior of the superstructure.

9. Cabinet vents in these hoods will be type 316 stainless steel, $1\frac{1}{2}$-in. tubing, downward oriented at 45°. They will be welded into the side walls 6 in. above the trough and behind the baffle. The hood manufacturer will supply and install stainless steel flexible tubing to connect the vents to the cabinets.

3.05 Radioisotope Hoods

1. The hood liner, including the work surface, will be 16 gage, type 304 stainless steel. It will be seamless, cove-welded construction with all welds ground and polished to a #4 satin finish. The flat panel areas will be a #2B mill finish.

2. The cup sink will be 3 in. by 6 in. and will be welded into the work surface.

3. The work surface will be reinforced with an industrial steel grating to allow a load distribution of 200 pounds per square foot without sag.

4. The duct collar will be welded into the hood liner.

5. Acid or solvent storage cabinet vents will be $1\frac{1}{2}$ in. stainless steel tubing welded into the back ledge of the work surface and will extend 2 in. above the work ledge.

6. Construction details of Part 3.03, this section, apply as applicable.

IF YOU ARE GOING TO INCLUDE ALARM SYSTEMS, VAV CONTROLS, OR SOMETHING SPECIAL IN OR ON YOUR HOODS, THEN YOU SHOULD GENERATE NEW PARAGRAPH NUMBERS HERE AND LIST YOUR CHANGES. THE PARAGRAPH NUMBERS FROM HERE TO THE END OF PART 3.07 WILL HAVE TO BE ADJUSTED ACCORDINGLY.

3.06 Fume Hood Performance Testing

1. The fume hood manufacturer, no later than 30 days after receipt of order, will provide to the owner a state-of-the-art

fume hood test facility meeting the requirements of SAMA Standard LF 10-1980.

2. The hood manufacturer will conduct the ASHRAE 110-1993 protocol of a 6-ft hood of similar design to the type specified.
3. The ASHRAE 110-1993 standard will have the following parameters for the purposes of this specification:
   a. The hood will be tested with simulated apparatus. This apparatus will consist of: two each 1 gallon, round paint cans, one each 12-in. by 12-in. by 12-in. cardboard box, three each 6-in. by 6-in. by 12-in. cardboard boxes. These items will be positioned from 6 to 10 in. behind the sash, randomly distributed, and supported off the work surface by 2-in. by 2-in. blocks.
   b. The hood will be tested with the sash full open and at 60, 80, and 100 fpm hood face velocities.
   c. The test gas will have a 4-lpm flow rate.
   d. The test will be conducted at the center position for the manikin only.
   e. Each test duration will be 5 minutes.
   f. Acceptable test results will be 4.0 AM .05 or better.
   g. At the conclusion of each 5-minute test there will be three rapid "walk-bys" at 1 ft behind the manikin. Each walk-by will be spaced 30 seconds apart. If there is a rise in test gas concentration, it cannot exceed 4.0 AM .20 and must return to 4.0 AM .05 within 15 seconds.
   h. There will be a minimum of three and a maximum of five people in the test room during the test procedure.
   i. Three representatives of the owner will witness the tests. The hood manufacturer is responsible for all expenses incurred by the owner's representatives to witness the testing.
   j. Failure to pass this test will be adequate cause for cancellation of the fume hood contract.

3.07 Post Installation Testing

A There will be a retesting of 50 percent of the fume hoods, in accordance with Part 3.06 (this section), after installation and building balance and prior to occupancy. These tests will not be conducted by the fume hood supplier but by an outside contractor with recognized credentials and at least 3 years experience in this field.

B This testing will be the responsibility of the mechanical contractor.

C For this post-installation testing a test rating of 4.0 AI .10 or better will be acceptable. (Note: "AI" signifies "As Installed.")

4.00 Execution

4.01 Inspection Prior to Installation The laboratory fume hood installer will inspect the work of the other trades and the installation conditions for acceptability. The installer will inform the general contractor and the architect of any problems that could jeopardize a proper installation.

4.02 Installation

A Install all fume hoods and associated countertops plumb and level.

B Touch up marred and/or abraded finished surfaces; clean components to post construction acceptance levels as specified by the architect. Remove all crating and packing materials to an area as designated by the general contractor.

C After the fume hoods have been installed and prior to touch-up, the hood installer will protect the work surfaces and vertical finished panels with plastic sheeting (minimum 5 mil thickness) or cardboard sheets taped in place.

Glossary

Air foil: Curved or angular member(s) at fume hood opening (face).

Air diffuser: A device used to distribute air from a ducting system into a room.

ASHRAE 110-1985: A Standard outlining the protocol to be used in the quantitative evaluation of fume hood performance.

Air volume: rate of air flow, normally expressed in cubic feet per minute (cfm).

Auxiliary air: Supply or secondary air delivered external to the superstructure of a fume hood to reduce room air consumption.

Baffle: A panel system located across the back interior portion of a fume hood to control air patterns and distribution.

Bypass hood: A hood with a compensating opening that maintains a relative constant exhaust volume through the hood regardless of sash position.

Cross draft: A flow of air that blows into or across the hood face.

Face: Front or access opening of the fume hood.

Face plane: The area formed by the external shell of the hood at the hood face.

Face velocity: The speed of the air moving into the hood through the face area normally expressed in feet per minute (fpm).

Hood diversity: The percentage of total hood population that operates simultaneously.

Liner: Interior lining material for the back, sides, top, and baffles of a fume hood superstructure.

Make-up (supply) air: Air needed to replace the air exhausted from the room by the fume hood(s).

Manometer: A device used to measure air pressure differential, usually calibrated in inches of water gage (wg).

Pitot tube: A device used for measuring the velocity pressure of air in a duct work system.

Sash: The movable transparent panel set in the fume hood face.

Side walls: The two end walls of the fume hood that enclose the hood chamber.

Static pressure: Air pressure in the fume hood chamber or ducting; should be negative but is expressed as a positive number in inches of water.

Superstructure: That portion of a fume hood supported by the base cabinets, nominally includes the working surface.

Variable air volume: A controlled mechanical system that varies the exhaust volume of the hood so as to maintain a constant face velocity.

Velometer: An instrument used to measure the velocity of air.

Vortex: Turbulence caused in the upper portion of a fume hood chamber by errant air currents.

Index